Lecture Notes in Mathematics

Edited by A. Dold and B. Eckmann

1307

Takafumi Murai

A Real Variable Method for the Cauchy Transform, and Analytic Capacity

Springer-Verlag

Berlin Heidelberg New York London Paris Tokyo

Author

Takafumi Murai
Department of Mathematics, College of General Education
Nagoya University
Nagoya, 464, Japan

Mathematics Subject Classification (1980): Primary 30C85; secondary 42A50

ISBN 3-540-19091-0 Springer-Verlag Berlin Heidelberg New York
ISBN 0-387-19091-0 Springer-Verlag New York Berlin Heidelberg

Printing and binding: Druckhaus Beltz, Hemsbach/Bergstr.
2146/3140-543210

PREFACE

The purpose of this lecture note is to study the Cauchy transform on curves and analytic capacity. For a compact set Γ in the complex plane $\mathbb{C}$, $H^\infty(\Gamma^c)$ denotes the Banach space of bounded analytic functions in $\mathbb{C}\cup\{\infty\}-\Gamma$ $(=\Gamma^c)$ with supremum norm $\|\cdot\|_{H^\infty}$. The analytic capacity of Γ is defined by

$$\gamma(\Gamma) = \sup\{|f'(\infty)|;\ \|f\|_{H^\infty} \leqq 1,\quad f\in H^\infty(\Gamma^c)\},$$

where $f'(\infty) = \lim_{z\to\infty} z(f(z)-f(\infty))$. We also define

$$\gamma_+(\Gamma) = \sup\{(1/2\pi)\int d\mu;\ \|C\mu\|_{H^\infty} \leqq 1,\quad C\mu\in H^\infty(\Gamma^c),\quad \mu \geqq 0\},$$

where

$$C\mu(z) = (1/2\pi i)\int 1/(\zeta-z)\, d\mu(\zeta) \qquad (z\notin(\text{the support of } \mu)).$$

We are concerned with estimating $\gamma(\cdot)$ and $\gamma_+(\cdot)$. To do this, compact sets having finite 1-dimension Hausdorff measure are critical. Hence we assume that Γ is a finite union of mutually disjoint smooth arcs. Let $|\cdot|$ denote the 1-dimension Hausdorff measure (the generalized length). Let $L^p(\Gamma)$ $(1\leqq p\leqq\infty)$ denote the L^p space of functions on Γ with respect to the length element $|dz|$, and let $L^1_w(\Gamma)$ denote the weak L^1 space of functions on Γ. Put

$$\rho(\Gamma) = \inf \gamma(E)/|E|,\quad \rho_+(\Gamma) = \inf \gamma_+(E)/|E|,$$

where the infimums are taken over all compact sets E in Γ. The Cauchy(-Hilbert) transform on Γ is defined by

$$H_\Gamma f(z) = (1/\pi)\ \text{p.v.} \int_\Gamma f(\zeta)/(\zeta-z)\ |d\zeta| \qquad (z\in\Gamma).$$

Then we see that

$$\rho_+(\Gamma) \leqq \rho(\Gamma) \leqq \text{Const}\ \rho_+(\Gamma)^{1/3},\quad \text{Const}\ \rho_+(\Gamma) \leqq 1/\|H_\Gamma\|_{L^1(\Gamma),L^1_w(\Gamma)} \leqq \text{Const}\ \rho_+(\Gamma),$$

where $\|H_\Gamma\|_{L^1(\Gamma),L^1_w(\Gamma)}$ is the norm of H_Γ as an operator from $L^1(\Gamma)$ to $L^1_w(\Gamma)$ (Theorem D). Hence the study of $\gamma(\Gamma)$ is closely related to the study of H_Γ.

Here is a history of the study of the Cauchy transform on Lipschitz graphs. According to Professor Igari, the L^2 boundedness of the Cauchy transform on Lipschitz graphs was first conjectured by Professor Zygmund in his lecture at Orsay in 1960's. Let $\Gamma = \{(x,A(x));\ x\in\mathbb{R}\}$, $a(x) = A'(x)$, where $\mathbb{R}$ is the real line. Let $C[a]$ denote the singular integral operator defined by a kernel $1/\{(x-y)+i(A(x)-A(y))\}$. Then the above conjecture means the following assertion: $C[a]$ is bounded (from $L^2(\mathbb{R})$ to itself) if $a\in L^\infty(\mathbb{R})$. The operator $C[a]$ is formally expanded in the following form: $(-\pi)H + \Sigma_{n=0}^\infty(-i)^n\, T_n[a]$, where H is the Hilbert transform and $T_n[a]$ is the singular integral operator defined by a kernel $(A(x)-A(x))^n/(x-y)^{n+1}$. In 1965, Calderón [3] showed that $T_1[a]$ is bounded if $a \in L^\infty(\mathbb{R})$ (Theorem A). This theorem is very important and closely related to the BMO($\mathbb{R}$) theory, where BMO($\mathbb{R}$) is the Banach space, modulo constants, of functions on $\mathbb{R}$ of bounded mean oscillation. Coifman-Meyer [8], [9] studied $T_n[a]$, Calderón [4] showed that $C[a]$ is bounded if $\|a\|_{L^\infty(\mathbb{R})}$ is sufficiently small, and consequently Coifman-McIntosh-Meyer [7] solved the above conjecture in the affirmative (Theorem B). David [17] studied H_Γ for continuous curves Γ. It is already known [44] that $\|C[a]\|_{L^2(\mathbb{R}),L^2(\mathbb{R})} \leq \mathrm{Const}(1 + \sqrt{\|a\|_{BMO(\mathbb{R})}})$ (Theorem C) and that the square root is best possible [18]. Jones-Semmes gives a simple proof of Theorem B by complex variable methods. (See Appendix II.)

As a first step of the study of H_Γ for discontinuous curves Γ, we begin with a review of the study of $C[a]$. In CHAP. I, 8 proofs of Theorem A will be given. Once this theorem is known, we can easily deduce Theorem B (cf. CHAP. II), and hence Theorem A is very important in the study of $C[a]$. As is easily seen, if $f, g \in L^2(\mathbb{R})$ have analytic extensions $f(z)$, $g(z)$ to the upper half plane (such that $\lim_{y\to\infty} f(iy) = \lim_{y\to\infty} g(iy) = 0$), then the Poisson extension of $(fg)(x)$ to the upper half plane is identical with $f(z)g(z)$. This simple property of analytic functions is essential in a proof of Theorem A by complex variable methods. We shall give, in CHAP. I, various interpretations of this property from the point of view of real analysis (cf. Coifman-Meyer-Stein [13]). These proofs are, of course, mutually very close, but each proof has proper applications and is interesting in itself.

In CHAP. II, we shall give the proofs of Theorems B and C by perturbation. Our method is an improvement of Calderón's perturbation [4] and David's perturbation [17]. Put

$$\sigma(C[a]) = \sup(1/|I|)\int_I |C[a](\chi_I f)(x)|\, dx,$$

where χ_I is the characteristic function of I and the supremum is taken over all intervals I and all real-valued functions f with $\|f\|_{L^\infty(\mathbb{R})} \leq 1$. This quantity is comparable to $\|C[a]\|_{L^\infty(\mathbb{R}),BMO(\mathbb{R})}$ and convenient for our perturbation. Considering a suitable Calderón-Zygmund decomposition of a primitive $A(x)$ of $a(x)$ on I, we obtain an a-priori estimate of $(1/|I|)\int_I |C[a](\chi_I f)(x)|\, dx$ by moderate graphs. (See the figure in § 2.2.) Repeating this argument infinitely many times and estimating infinitely many error terms, we see that the boundedness of $C[a]$ is consequently reduced to the boundedness of H. For the proof, Theorem A is necessary. We shall also give a proof of Theorem A by perturbation [45]. Tools which we use are only the Calderón-Zygmund decomposition and the covering lemma. For the proof of Theorem C, we put

$$\hat{\sigma}(C[a]) = \sup(1/|I|)\int_I |C[a](\chi_I f)(x)|^2\, f(x)\, dx,$$

where the supremum is taken over all intervals I and all real-valued functions f with $0 \leqq f \leqq 1$. Then $\sigma(C[a])^2 \leqq \text{Const}\, \hat{\sigma}(C[a])$. Since $\int_I C[a](\chi_I f)(x)f(x)\,dx = 0$, this quantity behaves like a linear functional of $a(x)$, and this gives an a-priori estimate better than $\sigma(C[a])$. Our method is not short but very simple, and this is applicable to various kernels.

In CHAP. III, we shall study H_Γ for discontinuous graphs Γ and shall compare $\gamma(\cdot)$ with integralgeometric quantities. We first give the proof of Theorem D. As is well-known, planar Cantor sets are useful to construct various examples (cf. Denjoy [23], Vitushkin [52]). Let $Q_0 = [0,1]\times[0,1]$ and let Q_n $(n \geqq 1)$ be the union of 4^n closed squares with sides of length 4^{-n} obtained from Q_{n-1} with each component of Q_{n-1} replaced by four squares in the four corners of the component. Put $Q_\infty = \bigcap_{n=0}^{\infty} Q_n$. Then $\gamma(Q_\infty) = 0$ and $|Q_\infty| > 0$ (Garnett [28]). This shows that two classes of null sets of $\gamma(\cdot)$ and $|\cdot|$ are different. We shall try to give grounds to this example. We may consider that Q_n is a graph. (See the figure in § 3.3.) Let $T_{s_1,\ldots,s_n}$ $(s_1,\ldots,s_n \in \mathbb{R})$ be the singular integral operator defined by a kernel

$$1/\{(x-y)+i(A_{s_1,\ldots,s_n}(x) - A_{s_1,\ldots,s_n}(y))\},$$

where $A_{s_1,\ldots,s_n}(x) = s_k$ $((k-1)/n \leqq x < k/n,\ 1 \leqq k \leqq n)$ and $A_{s_1,\ldots,s_n}(x) = 0$ $(x \in [0,1))$. Then we see that

$$\max\{\sigma(T_{s_1,\ldots,s_n});\ s_1,\ldots,s_n \in \mathbb{R}\}$$

is comparable to $\sqrt{\log(n+1)}$ (Theorem G), and, if we neglect constant multiples, an n-tuple $(s_1^0,\ldots,s_n^0)$ obtained from a graph $\{(x, A_{s_1^0,\ldots,s_n^0}(x));\ x \in [0,1)\}$ similar to Q_m (m = (the integral part of $(\log n)/4$)) is a solution of this extremal problem. Hence planar Cantor sets are worst curves in a sense. We shall also generalize Q_n. A segument $[0,1)$ is called a (thick) crank of degree 0 and a finite union Γ_n of segments parallel to the x-axis is called a (thick) crank of degree n, if Γ_n is obtained from a crank Γ_{n-1} of degree n-1 with each component J of Γ_{n-1} replaced by a finite number of segments $J_1,\ldots,J_{2^p}$ $(p=p(J))$ parallel to the x-axis such that $|J_k| = 2^{-p}|J|$, the distance between J_k and J is less than or equal to $2^{-p}|J|$ $(1 \leqq k \leqq 2^p)$ and the projections of these segments to $\mathbb{R}$ are mutually disjoint and contained in the projection of J. We shall show that, for any crank Γ of degree n, $\|H_\Gamma\|_{L^2(\Gamma),L^2(\Gamma)} \leqq \text{Const}\sqrt{n}$ and that this estimate is best possible (Theorem E). To prove this, we define n+1 singular integral operators $\{T_k\}_{k=0}^n$ such that $T_0 = (-\pi)H$, $\|\Sigma_{k=0}^n T_k\|_{L^2(\mathbb{R}),L^2(\mathbb{R})} = \|H_\Gamma\|_{L^2(\Gamma),L^2(\Gamma)}$ and $\{T_k\}_{k=0}^n$ are mutually almost orthogonal. Hence we see that the meaning of $\sqrt{n}$ is the central limit theorem.

We define integralgeometric quantities $Cr_\alpha(\cdot)$ $(0<\alpha<1)$ as follows. Let $D(z,r)$ be the open disk of center z and radius r. For a compact set E, $N_E(r,\theta)$ $(r>0, |\theta| \leqq \pi)$ denotes the (cardinal) number of elements of $E \cap L(r,\theta)$, where $L(r,\theta)$ is the straight line defined by the equation $x\cos\theta + y\sin\theta = r$. We put

$$Cr_\alpha(E) = \lim_{\varepsilon\to 0} Cr_\alpha^{(\varepsilon)}(E),$$

$$Cr_\alpha^{(\varepsilon)}(E) = \inf \int_{-\pi}^{\pi}\{\int_0^\infty N_{\partial\{\bigcup_{k=1}^n D(z_k,r_k)\}}(r,\theta)^\alpha\, dr\}\, d\theta \quad (\varepsilon>0),$$

where $\partial\{\bigcup_{k=1}^n D(z_k,r_k)\}$ is the boundary of $\bigcup_{k=1}^n D(z_k,r_k)$ and the infimum is taken over all finite coverings $\{D(z_k,r_k)\}_{k=1}^n$ of E with radii less than ε. Since $\gamma(E) \leqq \text{Const } Cr_1(E)$, it is interesting to compare $\gamma(\cdot)$ with $Cr_\alpha(\cdot)$ (cf. Marshall [37]). As an application of Theorem E, we shall show that, for $0<\alpha<1/2$, there exists a compact set E_α such that $\gamma(E_\alpha) = 1$ and $Cr_\alpha(E_\alpha) = 0$ (Theorem F). For the proof, we use a branching process. Let $\{X_n\}_{n=1}^\infty$ be a sequence of independent random variables on the standard probability space $([0,1),\mathcal{B},\text{Prob})$ such that $\text{Prob}(X_n = \pm 1) = 1/2$ $(n\geqq 1)$, and let $S_0 = 0$, $S_n = \Sigma_{k=1}^n X_k$ $(n\geqq 1)$. We define a Galton-Watson process $\{y_n\}_{n=0}^\infty$ by $y_0(x) = 1$, $y_n(x) = y_{n-1}(x) + S_{y_{n-1}(x)}(x)$ $(n\geqq 1)$. Then we see that, for $n\geqq 1$, there exists a crank Γ_n of degree n such that $Cr_\alpha(\Gamma_n)$ is comparable to $\Sigma_{k=0}^\infty k^\alpha\, \text{Prob}(y_n=k)$. This quantity is comparable to $1/n^{1-\alpha}$. Using the difference of order between $1/\sqrt{n}$ (the central limit theorem) and $1/n^{1-\alpha}$ (the Galton-Watson process), we construct the required set E_α.

I express my hearty thanks to Professors M.Ohtsuka, R.R.Coifman, P.W.Jones who gave me the chance to lecture during the academic year 1986-1987, and I am grateful to Professors S.Kakutani, T.Tamagawa, J.Garnett, S.Semmes, T.Steger, G.David, C.Bishop for their variable comments and suggestions. I especially express my appreciation to Professor W.H.J.Fuchs for his encouragement. I also thank to Mrs. Mel D. for typing the manuscript. This note is dedicated to the memory of my mother who died while I was staying at Yale University.

New Haven, July, 1987

CONTENTS

CHAPTER I. THE CALDERÓN COMMUTATOR
(8 PROOFS OF ITS BOUNDEDNESS)

§1.1. Calderón's Theorem (Calderón [3])

Let L^p $(1 \leq p \leq \infty)$ denote the L^p space on the real line $\mathbb{R}$ with respect to the 1-dimension Lebesgue measure $|\cdot|$. Its norm is denoted by $\|\cdot\|_p$. Let BMO denote the Banach space, modulo constants, of functions f on $\mathbb{R}$ such that $\|f\|_{BMO} = \sup(1/|I|) \int_I |f(x)-(f)_I| dx$ is finite, where the supremum is taken over all (finite) intervals I and $(f)_I$ is the mean of f over I. For $a \in L^\infty$, we define a kernel

$$(1.1) \quad T[a](x,y) = \{A(x) - A(y)\} /(x-y)^2,$$

where A is a primitive of a. We write simply by $T[a]$ the operator from L^2 to itself defined by the above kernel, i.e.,

$$(1.2) \quad T[a]f(x) = \lim_{\varepsilon \to 0} \int_{|x-y| > \varepsilon} T[a](x,y)f(y)dy.$$

Calderón showed

<u>Theorem A ([3])</u>. For any $f \in L^2$, $T[a]f(x)$ exists a.e.

$$(1.3) \quad \|a\|_\infty \leq \text{Const } \|T[a]\|_{2,2}$$

and

$$(1.4) \quad \|T[a]\|_{2,2} \leq \text{Const } \|a\|_\infty ,$$

where $\|T[a]\|_{2,2}$ is the norm of $T[a]$ (as an operator from L^2 to itself).

In §1.2, we show (1.3). In §1.3-1.11, we show various proofs of (1.4).

§1.2. Proof of (1.3) (Coifman-Rochberg-Weiss [15])

For a set $E \subset \mathbb{R}$, χ_E denotes the characteristic function of E. We put

$$\rho_\varepsilon(x) = \left|\int_{I_{+\varepsilon}} \{\int_{I_{-\varepsilon}} (A(s)-A(t))dt\}ds\right| \quad (x \in \mathbb{R}, \ \varepsilon > 0),$$

where $I_{+\varepsilon} = (x, x+\varepsilon)$, $I_{-\varepsilon} = (x-\varepsilon, x)$. Then $\lim_{\varepsilon \to 0} \rho_\varepsilon/\varepsilon^3 = \text{Const } |a|$ a.e. We have, for almost all x,

$$\rho_\varepsilon(x) = |\int_{I_{+\varepsilon}} [\int_{I_{-\varepsilon}} T[a](s,t)\{(s-x)^2 + 2(s-x)(x-t) + (x-t)^2\}\, dt\}\, ds|$$

$$\leq \int_{I_{+\varepsilon}} [(s-x)^2|T[a]\chi_{I_{-\varepsilon}}(s)| + 2|s-x|\,|T[a]\{(x-\cdot)\chi_{I_{-\varepsilon}}\}(s)|$$

$$+ |T[a]\{(x-\cdot)^2\chi_{I_{-\varepsilon}}\}(s)|]\, ds$$

$$\leq \text{Const}[\varepsilon^{5/2}\|T[a]\chi_{I_{-\varepsilon}}\|_2 + \varepsilon^{3/2}\|T[a]\{(x-\cdot)\chi_{I_{-\varepsilon}}\}\|_2$$

$$+ \varepsilon^{1/2}\|T[a]\{(x-\cdot)^2\chi_{I_{-\varepsilon}}\}\|_2]$$

$$\leq \text{Const}\,\|T[a]\|_{2,2}\{\varepsilon^{5/2}\|\chi_{I_{-\varepsilon}}\|_2 + \varepsilon^{3/2}\|(x-\cdot)\chi_{I_{-\varepsilon}}\|_2$$

$$+ \varepsilon^{1/2}\|(x-\cdot)^2\chi_{I_{-\varepsilon}}\|_2\} \leq \text{Const}\|T[a]\|_{2,2}\,\varepsilon^3,$$

and hence

$$|a| = \text{Const} \lim_{\varepsilon \to 0} \rho_\varepsilon/\varepsilon^3 \leq \text{Const}\|T[a]\|_{2,2} \quad \text{a.e.}$$

Thus we obtain (1.3).

§1.3. Area integral ([3])

In this section we show the proof of (1.4) by Calderón. Let C_0^∞ denote the totality of infinitely differentiable functions with compact support, $(\cdot,\cdot)$ denote the inner product and $\Upsilon_\varepsilon = \chi_{(-\varepsilon,\varepsilon)^c}$ $(\varepsilon > 0)$. Given real-valued functions a,f,g in C_0^∞ and $\varepsilon > 0$, we estimate

$$(T^\varepsilon[a]g,f) = \int_{-\infty}^{\infty} T^\varepsilon[a]g(x)f(x)dx,$$

where $T^\varepsilon[a]$ is an operator defined by a kernel $\Upsilon_\varepsilon(x-y)\,T[a](x,y)$. We may assume that $A(x) = \int_{-\infty}^{x} a(s)ds$. Then $A(x) = \int_{-\infty}^{\infty} e(x-s)a(s)ds$, where $e = \chi_{[0,\infty)}$. We have

$$(T^\varepsilon[a]g,f) = \int_{-\infty}^{\infty} a(s)[\int_{-\infty}^{\infty}\int_{-\infty}^{\infty} \frac{\Upsilon_\varepsilon(x-y)}{(x-y)^2}\{e(x-s)-e(y-s)\}g(y)f(x)dydx]ds.$$

Set

$$f_\pm(z) = \frac{1}{2\pi i}\int_{-\infty}^{\infty} \frac{f(x)}{x-z}\,dx \qquad \text{Im}\, z \begin{Bmatrix} > 0 \\ < 0 \end{Bmatrix}.$$

We denote also by $f_\pm(x)$ $(x \in \mathbb{R})$ the non-tangential limit of $f_\pm(z)$, respectively. We define analogously $g_\pm(z)$, $g_\pm(x)$. Then $f = f_+ - f_-$, $g = g_+ - g_-$, $\|f_\pm\|_2 \leq \|f\|_2$ and $\|g_\pm\|_2 \leq \|g\|_2$. Let

$$K_0(x,y,s) = \Upsilon_\varepsilon(x-y)\ \{e(x-s)-e(y-s)\}\ /(x-y)^2,$$

$$K_1^{\pm}\ (x,y,s) = \{e(x-s)-e(y-s)\}\ /(x-y \pm i\varepsilon)^2,$$

$$K_2(x,y,s) = \varepsilon/\{(x-s)^2 + (y-s)^2 + \varepsilon^2\}^{3/2}.$$

Then $|K_0(x,y,s) - K_1^{\pm}(x,y,s)| \leqq \text{Const}\ K_2(x,y,s)$. We have

$$|(T^\varepsilon[a]g,f)| = |\int_{-\infty}^{\infty} a(s)[\int_{-\infty}^{\infty} K_0(x,y,s)\{g_+(y) - g_-(y)\}\ f(x)dydx]\ ds|$$

$$\leqq |\int_{-\infty}^{\infty} a(s)[\int_{-\infty}^{\infty} K_1^+(x,y,s)g_+(y)f(x)dydx]\ ds|$$

$$+ |\int_{-\infty}^{\infty} a(s)[\int_{-\infty}^{\infty} K_1^-(x,y,s)g_-(y)f(x)dydx]\ ds|$$

$$+ \text{Const} \int_{-\infty}^{\infty} |a(s)|\ [\int_{-\infty}^{\infty} K_2(x,y,s)\ \{|g_+(y)| + |g_-(y)|\}\ |f(x)|\ dydx]ds$$

$$(= |\int_{-\infty}^{\infty} a(s)\ k_1^+(s)ds| + |\int_{-\infty}^{\infty} a(s)k_1^-(s)ds| + \text{Const} \int_{-\infty}^{\infty} |a(s)|k_2(s)ds, \text{say}).$$

We now estimate $k_1^{\pm}(s)$, $k_2(s)$. We have

$$k_1^+(s) = \int_{-\infty}^{\infty} f(x)\ \{\int_{-\infty}^{\infty} K_1^+(x,y,s)g_+(y)dy\}\ dx$$

$$= \int_{-\infty}^{\infty} f(x)\ \ \{e(x-s) \int_{-\infty}^{\infty} \frac{g_+(y)}{(x-y-i\varepsilon)^2}\ dy\ - \int_s^{\infty} \frac{g_+(y)}{(x-y-i\varepsilon)^2}\ dy\ \}\ dx$$

$$= -i \int_{-\infty}^{\infty} f(x)\ [\int_0^{\infty} g_+(s+it)/\{(x-i\varepsilon)-(s+it)\}^2\ dt]\ dx$$

$$= -i \int_0^{\infty} g_+(s+it)\ [\int_{-\infty}^{\infty} f(x)/\{(x-i\varepsilon)-(s+it)\}^2\ dx]\ dt$$

$$= 2\pi \int_0^{\infty} f_+'(s+i(t+\varepsilon))g_+(s+it)\ dt.$$

Let

$$F(z) = -i \int_0^{\infty} f_+'(z+i(t+\varepsilon))\ g_+(z+it)\ dt \qquad (z \in U),$$

where $U = \{(x,y);\ x \in \mathbb{R},\ y > 0\}$. Then F is analytic in U and the non-tangential limit $F(s)$ equals $(1/2\pi i)\ k_1^+(s)$. Here is a main lemma necessary for the proof of (1.4). Let $P_y(x)$ be the Poisson kernel, i.e., $P_y(x) = y/\{\pi(x^2+y^2)\}$. For a differentiable function $v(x,y)$ in U, we write $|\nabla v(x,y)| = \{|\partial v/\partial x|^2 + |\partial v/\partial y|^2\}^{1/2}$.

Lemma 1.1 ([3]). For $v \in L^1$, we define

$$A(v)(x) = \{\iint_{\Delta(x)} |\nabla v(\xi,\eta)|^2\ d\xi\ d\eta\}^{1/2} \qquad (x \in \mathbb{R}),$$

where $v(\xi,\eta) = P_\eta * v(\xi)$ and $\Delta(x) = \{(\xi,\eta);\ |\xi - x| < \eta\}$. Then $\|v\|_1 \leqq \text{Const}\|A(v)\|_1$.

Once this lemma is known, (1.4) is deduced as follows. Since $F'(z) = f_+'(z + i\varepsilon)g_+(z)$, we have $A(F)(s) \leqq A(f_+)(s)m(g_+)(s) \leqq \text{Const}\ A(f_+)(s)\ \mathcal{M} g_+(s)$, where $m(g_+)(s) = \sup\{|g_+(\xi,\eta)|;(\xi,\eta) \in \Delta(x)\}$ and $\mathcal{M}$ is the non-centered maximal operator (Journé [35, p.6]). (See Lemma 2.3.) We have $\|\mathcal{M}g_+\|_2 \leqq \text{Const}\|g_+\|_2$. Green's formula shows that $\|A(f_+)\|_2 = \text{Const}\|f_+\|_2$. Thus we have, by Lemma 1.1,

$$\begin{aligned}|\textstyle\int_{-\infty}^{\infty} a(s)k_1^+(s)ds| &\leqq 2\pi\ \|a\|_\infty\ \|F\|_1 \leqq \text{Const}\|a\|_\infty\ \|A(F)\|_1\\ &\leqq \text{Const}\ \|a\|_\infty\ \|A(f_+)\|_2\ \|m(g_+)\|_2 \leqq \text{Const}\ \|a\|_\infty\ \|f_+\|_2\ \|g_+\|_2\\ &\leqq \text{Const}\ \|a\|_\infty\ \|f\|_2\ \|g\|_2 .\end{aligned}$$

In the same manner, we have $|\int_{-\infty}^{\infty} a(s)k_1^+(s)ds| \leqq \text{Const}\ \|a\|_\infty\ \|f\|_2\ \|g\|_2$. We have

$$\begin{aligned}k_2(s) &\leqq \int_{-\infty}^{\infty}\frac{\varepsilon}{(x-s)^2+\varepsilon^2}\,|f(x)|\ \Big[\int_{-\infty}^{\infty}\frac{\sqrt{(x-s)^2+\varepsilon^2}}{(x-s)^2+(y-s)^2+\varepsilon^2}\{|g_+(y)|+|g_-(y)|\}dy\Big]dx\\ &\leqq \text{Const}\ \mathcal{M}f(s)\ \{\mathcal{M}g_+(s) + \mathcal{M}g_-(s)\},\end{aligned}$$

and hence

$$\int_{-\infty}^{\infty} |a(s)|k_2(s)\,ds \leqq \text{Const}\ \|a\|_\infty\ \|f\|_2\ \|g\|_2 .$$

Consequently $|(T^\varepsilon[a]g,f)| \leqq \text{Const}\ \|a\|_\infty\ \|f\|_2\ \|g\|_2$. Since $f,g \in C_0^\infty$, $\varepsilon > 0$ are arbitrary, we have (1.4) for $a \in C_0^\infty$. In the general case, we can deduce (1.4) from the boundedness of maximal operators $T^*[b]$ $(b \in C_0^\infty)$ and Fatou's lemma. (See Lemma 2.5.)

§1.4. Good λ inequalities ([2], [26], [48])

In this section we give the proof of Lemma 1.1 by the so-called "good λ inequalities". We put $\tilde{m}(v)(x) = \sup\{|v(x,y)|;\ y > 0\}$. Fixing a sufficiently large τ, we prove

$$\begin{aligned}(1.5)\qquad &|x;\ \tilde{m}(v)(x) > \tau\lambda\ ,\ A(v)(x) \leqq \lambda/\tau\ |\\ &\leqq (\text{Const}/\tau^2)\ |x;\ \tilde{m}(x) > \lambda| \qquad (\lambda > 0).\end{aligned}$$

Let $W(\lambda) = \{x;\ \tilde{m}(x) > \lambda\}$, $\delta(\lambda) = |W(\lambda)|$. Then we can write $W(\lambda) = \bigcup_{k=1}^\infty I_k$ with a sequence $\mathcal{M}_\lambda = \{I_k\}$ of mutually disjoint open intervals. It is sufficient to show that, for each $I \in \mathcal{M}_\lambda$,

$$(1.6)\qquad |E\ | \leqq (\text{Const}/\tau^2)\ |I|,$$

where $E = \{x \in I; \tilde{m}(v)(x) > \tau\lambda, A(v)(x) \leqq \lambda/\tau\}$. To do this we may assume that $A(v)(\xi) \leqq \lambda/\tau$ for some $\xi \in I$; otherwise $E = \emptyset$. Since $A(v)(\xi) \leqq \lambda/\tau$, we have, for any $x \in I$, $y \geqq 2|I|$,

(1.7) $$|v(\alpha,y) - v(x,y)| \leqq \text{Const } A(v)(\xi) \leqq \text{Const } \lambda/\tau ,$$

where α is the left endpoint of I. We choose τ large enough so that the last quantity in (1.7) is less than λ. Since $\tilde{m}(v)(\alpha) \leqq \lambda$, we have $|v(x,y)| \leqq 2\lambda$ $(x \in I, y \geqq 2|I|)$. Hence, for any $x \in E$, there exists $0 < y_x < 2|I|$ such that $|v(x,y_x)| = \sup\{|v(x,y)|; y > y_x\} = \tau\lambda$. Let $J(x) = (x - (y_x/5), x + (y_x/5))$, $\tilde{J}(x) = \{(\xi,y_x); |\xi - x| < y_x/10\}$ $(x \in E)$. Then, for any $(\xi,y_x) \in \tilde{J}(x)$, we have $|v(\xi,y_x)| \geqq |v(x,y_x)| - \text{Const } A(v)(x) \geqq \tau\lambda - \text{Const } \lambda/\tau \geqq \tau\lambda/2$. There exist a finite number of mutually disjoint intervals $\{J(x_\mu)\}$ such that $|E| \leqq 5 \Sigma |J(x_\mu)|$. (See §2.2.) Let $R = Q_0 \cap \cup \tilde{\Delta}(x_\mu)$, where $Q_0 = \{(\xi,\eta); \xi \in I, 0 < \eta < 2|I|\}$, $\tilde{\Delta}(x_\mu) = \{(\xi,\eta); |\xi - x_\mu| < \eta/10, \eta > y_{x_\mu}\}$. Green's formula shows that

(1.8) $$\int_{\partial R} \left\{ \frac{\partial\eta}{\partial n} |v|^2 - \eta \frac{\partial |v|^2}{\partial n} \right\} ds = \text{Const} \iint_R \eta |\nabla v|^2 \, d\xi \, d\eta,$$

where $\partial/\partial n$ is the inner normal derivative and ds is the length element. Let $A_R(v)(x) = \{\iint_{\Delta^*(x) \cap R} |\nabla v|^2 \, d\xi \, d\eta\}^{1/2}$, where $\Delta^*(x) = \{(\xi,\eta); |\xi - x| < \eta/10\}$. Then a geometric observation shows that $A_R(v)(x) \leqq A(v)(x_\nu) \leqq \lambda/\tau$, where x_ν is a point which is nearest to x in $\{x_\mu\}$. Hence the right-hand side of (1.8) is dominated by:

$$\text{Const} \int_I A_R(v)(x)^2 dx \leqq \text{Const}(\lambda/\tau)^2 |I| \leqq \text{Const } \lambda^2 |I|.$$

We divide ∂R into the following three parts: $\partial R_0 = \partial R \cap \cup \tilde{J}(x_\mu)$, $\partial R_1 = \{(\xi,\eta); \xi \in I, \eta = 2|I|\}$, $\partial R_2 = \partial R - (\partial R_0 \cup \partial R_1)$. Note that $\eta|\nabla v(\xi,\eta)| \leqq \text{Const } \lambda/\tau$ on ∂R. By the definition of y_x $(x \in E)$, we have, for any $(\xi,\eta) \in \partial R$, $|v(\xi,\eta)| \leqq \tau\lambda + \text{Const } \lambda/\tau \leqq \text{Const } \tau\lambda$. Thus

$$\left|\int_{\partial R} \eta \frac{\partial |v|^2}{\partial n} ds\right| \leqq \text{Const} \int_{\partial R} \eta |\nabla v| |v| \, ds$$

$$\leqq \text{Const } (\lambda/\tau) \tau\lambda \int_{\partial R} ds \leqq \text{Const } \lambda^2 |I|.$$

Since $|v(\xi,\eta)| \leqq \text{Const } \lambda$ on ∂R_1, we have $\left|\int_{\partial R_1} \frac{\partial\eta}{\partial n} |v|^2 ds\right| \leqq \text{Const } \lambda^2 |I|$. These estimates yield that $\int_{\partial R_0 \cup \partial R_2} \frac{\partial\eta}{\partial n} |v|^2 ds \leqq \text{Const } \lambda^2 |I|$. Since $\partial\eta/\partial n \geqq 0$ on ∂R_2, $\partial\eta/\partial n = 1$ on ∂R_0 and $|v(\xi,\eta)| \geqq \tau\lambda/2$ on ∂R_0, we have

$$\tau^2\lambda^2|E| \leq \mathrm{Const}\int_{\partial R_0} \frac{\partial\eta}{\partial n}|v|^2 ds \leq \mathrm{Const}\int_{\partial R_0 \cup \partial R_2} \frac{\partial\eta}{\partial n}|v|^2 ds \leq \mathrm{Const}\,\lambda^2\,|I|,$$

which shows (1.6). Consequently (1.5) holds.

By (1.5), we have, with a constant C_0,

$$(1.9) \qquad \delta(\tau\lambda) \leq \tilde{\delta}(\lambda/\tau) + (C_0/\tau^2)\,\delta(\lambda),$$

where $\tilde{\delta}(\lambda) = |x;\ A(v)(x) > \lambda|$. We now choose $\tau = 2\,C_0$ and integrate each quantity in (1.9) by $d\lambda$ from 0 to infinity. Then we obtain $\|\tilde{m}(v)\|_1 \leq \mathrm{Const}\,\|A(v)\|_1$, which gives $\|v\|_1 \leq \mathrm{Const}\,\|A(v)\|_1$. This completes the proof of Lemma 1.1.

§1.5. BMO (Fefferman-Stein [27])

Theorem A is closely related to the theory of BMO [27]. In this section, we show the proof of Theorem A by Fefferman-Stein. We say that a non-negative measure $d\mu(x,y)$ in U is a Carleson measure with constant B if

$$\iint_{I\times(0,|I|)} d\mu(x,y) \leq B\,|I|$$

for any interval $I \subset R$. The following two facts are elementary.

Lemma 1.2 ([27]). Let $a \in$ BMO. Then $y|\nabla a(x,y)|^2\,dx\,dy$ is a Carleson measure with constant $\mathrm{Const}\,\|a\|^2_{BMO}$, where $a(x,y) = P_y * a(x)$.

Proof. Given an interval I, we put

$$a^{(1)}(x) = (a(x) - (a)_I)\,\chi_{I^*}(x),\quad a^{(2)}(x) = (a(x) - (a)_I)\,\chi_{I^{*c}}(x),$$

where $(a)_I = (1/|I|)\int_I a(y)dy$ and I^* is the double of I, i.e., the (open) interval of the same midpoint as I and of length $2|I|$. Then

$$a(x,y) = P_y * a^{(1)}(x) + P_y * a^{(2)}(x) + (a)_I$$
$$(= a^{(1)}(x,y) + a^{(2)}(x,y) + (a)_I,\ \text{say}).$$

John-Nirenberg's inequality [32] shows that $\|a^{(1)}\|_2 \leq \mathrm{Const}\,\|a\|_{BMO}\sqrt{|I|}$. (See Lemma 2.5.) Hence we have, with $\hat{I} = I \times (0,|I|)$,

$$\iint_{\hat{I}} y\,|\nabla a^{(1)}(x,y)|^2\,dx\,dy \leq \iint_U y\,|\nabla a^{(1)}(x,y)|^2\,dx\,dy$$
$$= \mathrm{Const}\,\|a^{(1)}\|_2^2 \leq \mathrm{Const}\,\|a\|^2_{BMO}\,|I|.$$

Note that $|(a)_{I_j} - (a)_I| \leq \mathrm{Const}\ j\|a\|_{BMO}$ $(j \geq 1)$, where I_j is the interval of the same midpoint as I and of length $2^j|I|$. We have, for $(x,y) \in \hat{I}$

$$|\nabla a^{(2)}(x,y)| \leq \text{Const} \int_{I^{*c}} \frac{1}{(x-s)^2} |a^{(2)}(s)|\,ds$$

$$\leq \text{Const} \sum_{j=1}^{\infty} |I_j|^{-2} \int_{I_{j+1}-I_j} |a(y)-(a)_I|\,ds$$

$$\leq \text{Const} \sum_{j=1}^{\infty} |I_j|^{-2} |I_{j+1}| \{\|a\|_{BMO} + |(a)_{I_j} - (a)_I|\}$$

$$\leq \text{Const}(\sum_{j=1}^{\infty} j\, 2^{-j}) \|a\|_{BMO}/|I| ,$$

and hence

$$\iint_{\hat{I}} y\, |\nabla a^{(2)}(x,y)|^2\, dx\, dy \leq \text{Const}(\|a\|_{BMO}/|I|)^2 \iint_{\hat{I}} y\, dx\, dy$$

$$\leq \text{Const}\ \|a\|_{BMO}^2\, |I| .$$

Thus

$$\iint_{\hat{I}} y\ |\nabla a(x,y)|^2\, dx\, dy \leq \text{Const}\ \{\iint_{\hat{I}} y\, |\nabla a^{(1)}(x,y)|^2\, dx\, dy$$

$$+ \iint_{\hat{I}} y\ |\nabla a^{(2)}(x,y)|^2\, dx\, dy\} \leq \text{Const}\ \|a\|_{BMO}^2\, |I| .$$ Q.E.D.

Lemma 1.3 ([35, p. 85]). Let $d\mu(x,y)$ be a Carleson measure with constant B. Then, for any $f \in L^2$,

$$\iint_U |f(x,y)|^2\, d\mu(x,y) \leq \text{Const}\ B\|f\|_2^2 \qquad (f(x,y) = P_y * f(x)).$$

Proof. Let $W(\lambda) = \{(x,y) \in U;\ |f(x,y)| > \lambda\}$, $\delta(\lambda) = \iint_{W(\lambda)} d\mu(x,y)$ ($\lambda > 0$). Then the left-hand side of our lemma is dominated by

Const $\int_0^\infty \lambda\delta(\lambda)d\lambda$. If $(x,y) \in W(\lambda)$, then
$\lambda \leq \sup\{|f(\xi,\eta)|;\ |x-\xi| < \eta\} \leq C\mathcal{M}f(x)$ for some constant C. Hence $W(\lambda)$ is contained in $W_0(\lambda) = \cup\, I \times (0,|I|)$, where the union is taken over all components I of $\{x;\ \mathcal{M}f(x) > C\lambda\}$. Thus

$$\delta(\lambda) \leq \iint_{W_0(\lambda)} d\mu(x,y) \leq B|x;\ \mathcal{M}f(x) > C\lambda| ,$$

which gives

$$\int_0^\infty \lambda\, \delta(\lambda)d\lambda \leq B \int_0^\infty \lambda|x;\ \mathcal{M}f(x) > C\lambda|\,d\lambda$$

$$\leq \text{Const}\ B\ \|\mathcal{M}f\|_2^2 \leq \text{Const}\ B\ \|f\|_2^2 .$$ Q.E.D.

We now prove Theorem A. The Hilbert transform H is defined by

$$Hf(x) = \frac{1}{\pi} \lim_{\varepsilon \to 0} \int_{|s-x| > \varepsilon} \frac{f(s)}{s-x}\, ds .$$

For $a, f \in C_0^\infty$, we have

(1.10) $T[a]f(x) = -\pi\, H(af)(x) + \pi[A,H]f'(x)$,

where $[A,H]f' = A(Hf') - H(Af')$. Since $\|H(af)\|_2 \leqq \|a\|_\infty \|f\|_2$, it is sufficient to show that $\|[A,H]f'\|_2 \leqq \text{Const}\,\|a\|_\infty \|f\|_2$; we will prove a better inequality.

(1.11) $\|[A,H]f'\|_2 \leqq \text{Const}\,\|a\|_{BMO}\,\|f\|_2$.

Without loss of generality we may assume that a, f are real-valued. We have, for any real-valued function $g \in C_0^\infty$,

$$([A,H]f',g) = \int_{-\infty}^{\infty} [A,H]f'(x)g(x)dx = (A, Hf'\cdot g + f'Hg)$$

$$= 4\,\text{Im}(A, f_+'\, g_+) = 4\,\text{Im}(A,F') = -4\,\text{Im}(a,F),$$

where

(1.12) $F(x) = \int_{-\infty}^{x} f_+'(s)g_+(s)ds = -i\int_0^\infty f_+'(x+is)g_+(x+is)ds.$

Let $a(x,y) = P_y * a(x,y)$, $F(x,y) = P_y * F(x)$. Since $f_+'(z)$, $g_+(z)$ are analytic in U, we have $\frac{\partial F}{\partial x}(x,y) = f_+'(x+iy)g_+(x+iy)$. Thus Lemmas 1.2, 1.3 and Parseval's formula yield that

$$|(a,F)| = \text{Const}\,\Big|\iint_U y\,\frac{\partial a}{\partial x}(x,y)\,\frac{\partial F}{\partial x}(x,y)\,dx\,dy\Big|$$

$$= \text{Const}\,\Big|\iint_U y\,\frac{\partial a}{\partial x}(x,y)f_+'(x+iy)g_+(x+iy)\,dx\,dy\Big|$$

$$\leqq \text{Const}\,\{\iint_U y|f_+'(x+iy)|^2\,dx\,dy\}^{1/2}\,\{\iint_U y\,|\nabla a(x,y)|^2|g_+(x+iy)|^2 dx\,dy\}^{1/2}$$

$$\leqq \text{Const}\,\|f_+\|_2\,\|a\|_{BMO}\,\|g_+\|_2 \leqq \text{Const}\,\|a\|_{BMO}\,\|f\|_2\,\|g\|_2 .$$

This completes the proof of Theorem A.

Fefferman-Stein [27] showed also the following inequality, which is essentially same as (1.11).

Lemma 1.4 ([27]). Let $a \in BMO$. Then $\|[a,H]\|_{2,2} \leqq \text{Const}\,\|a\|_{BMO}$.

Proof. Without loss of generality we may assume that a is real-valued. We have, for any real-valued functions $f,g \in C_0^\infty$,

$$([a,H]f,g) = (a, Hf\cdot g + fHg) = -4\,\text{Im}\,(a, f_+g_+).$$

Let $G(x) = f_+(x)g_+(x)$. Then Parseval's formula shows that, with $G(x,y) = P_y * G(x)$, $a(x,y) = P_y * a(x)$,

$$|(a,f_+g_+)| = |(a,G)| = \text{Const}\ \left|\iint_U y\ \frac{\partial a}{\partial x}(x,y)\ \frac{\partial G}{\partial x}(x,y)\ dx\,dy\right|$$

$$\leqq \text{Const}\ \{\iint_U y\ |\nabla a|^2\ |G|\ dx\,dy\}^{1/2}\ \{\iint_U y\ |\nabla G|^2\ |G|^{-1}\ dx\,dy\}^{1/2}.$$

Since $\log|G(x,y)|$ is subharmonic in U,

$$\Delta \log|G| = (\Delta|G| - \frac{|\nabla|G||^2}{|G|})\ \frac{1}{|G|} \geqq 0,$$

and hence

$$\frac{|\nabla G|^2}{|G|} = \Delta|G| + \frac{|\nabla|G||^2}{|G|} \leqq 2\,\Delta|G|.$$

This shows that

$$\iint_U y\ |\nabla G|^2\ |G|^{-1}\ dx\,dy \leqq 2\ \iint_U y\ \Delta|G|\ dx\,dy = \text{Const}\ \|G\|_1.$$

Since $|G(x,y)|^{1/2}$ is subharmonic in U, we have $|G(x,y)| \leqq P_y * (|G|^{1/2})(x)^2$. Hence Lemmas 1.2 and 1.3 yield that

$$\iint_U y\ |\nabla a(x,y)|^2 |G(x,y)|\ dx\,dy \leqq \iint_U y\ |\nabla a(x,y)|^2\ P_y * (|G|^{1/2})(x)^2 dx\,dy$$

$$\leqq \text{Const}\ \|a\|_{BMO}^2\ \|G\|_1.$$

Consequently, we have

$$|([a,H]f,g)| \leqq \text{Const}\ \|a\|_{BMO}\ \|G\|_1 \leqq \text{Const}\ \|a\|_{BMO}\ \|f\|_2\ \|g\|_2.$$ Q.E.D.

§1.6. The Coifman-Meyer expression (Coifman-Meyer [8])

It is important to understand Theorem A from the point of view of real analysis. Coifman-Rochberg-Weiss [15] showed Lemma 1.4 without using analytic functions. Coifman-Meyer gave the following expression.

Lemma 1.5 ([8]). $[A,H]f'(x)$

$$= -\ \text{Const} \int_{-\infty}^{\infty} [a_{-s},H]f_s(x)/(1+s^2)ds \quad (a \in BMO,\ f \in C_0^\infty),$$

where $a_s = k_s * a$, $f_s = k_s * f$, $k_s(x) = \Xi_s/|x|^{1+is}$ and $\Xi_s = \Gamma((1+is)/2)/\ \{\Gamma(-is/2)\ \pi^{is}\}$.

Proof. We have, for $a, f \in C_0^\infty$,

$$[A,H]f'(x) = \text{Const}\ i \int_{-\infty}^{\infty}\int_{-\infty}^{\infty} e^{i(\xi+\eta)x}\ \{\text{sign}\,\eta - \text{sign}(\xi+\eta)\}\ \frac{\hat{a}(\xi)}{i\xi}\ i\eta\hat{f}(\eta)d\xi\,d\eta\ ,$$

where $\hat{a}$, $\hat{f}$ are the Fourier transform of a, f, respectively. Note that

$\{\text{sign}\,\eta - \text{sign}(\xi+\eta)\}\,(\eta/\xi) = -\{\text{sign}\,\eta - \text{sign}(\xi+\eta)\}\,\chi_{(0,1)}(|\eta/\xi|)$. Since

$$|\eta/\xi| = \text{Const}\int_{-\infty}^{\infty} |\eta/\xi|^{is}/(1+s^2)\,ds \qquad (|\eta/\xi| \leqq 1),$$

$\hat{a}_{-s}(\xi) = \hat{k}_{-s}(\xi)\hat{a}(\xi) = |\xi|^{-is}\,\hat{a}(\xi)$ and $\hat{f}_s(\eta) = |\eta|^{is}\,\hat{f}(\eta)$, we have

$$[A,H]f'(x) = -\text{Const}\int_{-\infty}^{\infty}[i\int_{-\infty}^{\infty}\int_{-\infty}^{\infty} e^{i(\xi+\eta)x}\{\text{sign}\,\eta - \text{sign}(\xi+\eta)\}$$
$$\hat{a}_{-s}(\xi)\,\hat{f}_s(\eta)\,d\xi\,d\eta]/(1+s^2)\,ds = -\text{Const}\int_{-\infty}^{\infty}[a_{-s},H]f_s(x)/(1+s^2)\,ds.$$

(In the case of $a \in BMO$, $f \in C_0^\infty$, it is necessary to show the convergence of the quantity in the right-hand side of Lemma 1.5. This will be shown later in the proof of Theorem A.) Q.E.D.

Here is another lemma necessary for the proof of Theorem A.

Lemma 1.6 ([8]). $\|a_s\|_{BMO} \leqq \text{Const}(1+|s|^{3/4})\,\|a\|_{BMO}$.

Proof. Without loss of generality we may assume that $s > 0$. We put $a^{(1)} = (a-(a)_I)\,\chi_{I^*}$, $a^{(2)} = (a-(a)_I)\,\chi_{I^{*c}}$. (See Lemma 1.2.) Then $a_s = a_s^{(1)} + a_s^{(2)}$, where $a_s^{(j)} = k_s * a^{(j)}$ $(j = 1,2)$. John-Nirenberg's inequality shows that $\|a^{(1)}\|_2 \leqq \text{Const}\|a\|_{BMO}\sqrt{|I|}$, and hence

$$\int_I |a_s^{(1)}(x)|\,dx \leqq \|a_s^{(1)}\|_2\sqrt{|I|} = \|a^{(1)}\|_2\sqrt{|I|} \leqq \text{Const}\,\|a\|_{BMO}\,|I|.$$

Note that $|\Xi_s| \leqq \text{Const}(1+\sqrt{s})$. In the same manner as in Lemma 1.2, we have, with $x_0 =$ (the midpoint of I),

$$\int_I |a^{(2)}(x) - a^{(2)}(x_0)|\,dx$$
$$= |\Xi_s|\int_I \Big|\int_{-\infty}^{\infty}\Big\{\frac{1}{|x-y|^{1+is}} - \frac{1}{|x_0-y|^{1+is}}\Big\}\,a^{(2)}(y)dy\Big|\,dx$$
$$\leqq \text{Const}\,\{|\Xi_s|\,(1+s^{1/4})\}\int_I |x-x_0|^{1/4}\Big\{\int_{I^{*c}}\frac{1}{|x_0-y|^{5/2}}\,|a(y)-(a)_I|dy\Big\}\,dx$$
$$\leqq \text{Const}\,(1+s^{3/4})\,\|a\|_{BMO}\,|I|.$$

Thus we obtain

$$(|a_s - (a_s)_I|)_I \leqq 2(|a_s - a_s^{(2)}(x_0)|)_I \leqq \text{Const}\,(1+s^{3/4})\,\|a\|_{BMO},$$

which gives the required inequality. Q.E.D.

Theorem A is deduced from Lemmas 1.4-1.6 as follows. Inequality (1.10) shows that it is sufficient to show that $\|[A,H]f'\|_2 \leqq \text{Const}\,\|a\|_\infty\|f\|_2$. Lemmas 1.4-1.6

yield that

$$\|[A,H]f'\|_2 \leqq \text{Const} \int_{-\infty}^{\infty} \|[a_s,H]f_s\|_2/(1+s^2)\, ds$$

$$\leqq \text{Const} \int_{-\infty}^{\infty} \|[a_s,H]\|_{2,2} \|f_s\|_2/(1+s^2)\, ds$$

$$\leqq \text{Const} \|f\|_2 \int_{-\infty}^{\infty} \|a_s\|_{BMO}/(1+s^2)\, ds$$

$$\leqq \text{Const} \|a\|_{BMO} \|f\|_2 \leqq \text{Const} \|a\|_\infty \|f\|_2 .$$

§1.7. A tent space (Coifman-Meyer-Stein [13], [14])

The essential part in the sections 1.3 and 1.5 is the proof of the inequality: $\|F\|_1 \leqq \text{Const} \|f\|_2 \|g\|_2$ $(f, g \in L^2)$, where $F(x) = \int_{-\infty}^{x} f_+'(s)g_+(s)\, ds$ $(= -i\int_0^\infty f_+'(x+is)g_+(x+is)ds)$. Let R_s $(s \in \mathbb{R})$ denote the operator defined by $R_s h = \psi_s * h$, where $\psi_s(x) = s^2x/(x^2+s^2)^2$. Then we have

$F(x) = \text{Const} \int_0^\infty R_s h_s(x)\, ds/s$, where $h_s(y) = s\, f_+'(y+is)g_+(y+is)$. From this point of view, Coifman-Meyer-Stein introduces tent spaces and generalizes the above inequality. As seen in the proof of the Tb theorem (David-Journé-Semmes [20]), tent spaces are very useful. The following theorem is a special case of Coifman-Meyer-Stein's theorem; we rewrite their theorem so that only the proof of Theorem A can be given.

Theorem 1.7 ([13]). Let T be the Banach space of functions $h(y,s)$ in U with norm $\|h\|_T = \|S(h)\|_1$, where $S(h)(x) = \{\iint_{\Delta(x)} |h(y,s)|^2\, dyds/s^2\}^{1/2}$ and $\Delta(x) = \{(y,s);\ |y-x| < s\}$. For $h \in T$, we put $R(h)(x) = \int_0^\infty R_s h_s(x) ds/s$, where $h_s(y) = h(y,s)$. Then $\|R(h)\|_1 \leqq \text{Const} \|h\|_T$.

Theorem A immediately follows from this inequality, since

$$\|sf_+'(y+is)g_+(y+is)\|_T \leqq \|A(f_+)m(g_+)\|_1 \leqq \text{Const} \|f\|_2 \|g\|_2 .$$

Here are two lemmas necessary for the proof. For an interval I, we write

$$\hat{I} = \{(y,s);\ 0 < s < \text{dis}(y,I^c)\} \quad (\text{dis}(\cdot,\cdot) \text{ is the distance}).$$

For an open set $\Omega \subset \mathbb{R}$, we write $\hat{\Omega} = \cup\, \hat{I}$, where the union is taken over all components I of Ω. We say that $p \in T$ is a T-atom if, for some interval I,

$$(1.13) \quad \text{supp}(p) \subset \hat{I}, \quad \iint_{\hat{I}} |p(y,s)|^2 \frac{dy\, ds}{s} \leqq 1/|I|,$$

where $\text{supp}(\cdot)$ is the support.

Lemma 1.8 ([13]). For any T-atom p, $\|R(p)\|_1 \leqq \text{Const}$.

Proof. Let p be a T-atom and let I be an interval satisfying (1.13). Then, for any $b \in L^\infty$ with $\|b\|_\infty \leq 1$, $|(R(p),\bar{b})| \leq |(R(p),\bar{b}_1)| + |(R(p),\bar{b}_2)|$ where $b_1 = b\chi_{I^*}$ and $b_2 = b\chi_{I^{*c}}$ (I^* is the double of I). Since $|R_s b_2(y)| \leq \text{Const}\ s/|I|$ ($(y,s) \in \hat{I}$), we have

$$|(R(p),\bar{b}_2)| = |\iint_{\hat{I}} p(y,s)\, R_s b_2(y) \frac{dy\,ds}{s}|$$

$$\leq (\text{Const}/|I|) \iint_{\hat{I}} |p(y,s)|\ dy\,ds$$

$$\leq (\text{Const}/|I|) \{\iint_{\hat{I}} |p(y,s)|^2 \frac{dy\,ds}{s}\}^{1/2} \{\iint_{\hat{I}} s\,dy\,ds\}^{1/2} \leq \text{Const}.$$

We have

$$|(R(p),\bar{b}_1)| \leq \iint_{\hat{I}} |p(y,s)||R_s b_1(y)| \frac{dy\,ds}{s}$$

$$\leq \{\iint_{\hat{I}} |p(y,s)|^2 \frac{dy\,ds}{s}\}^{1/2} \{\iint_{\hat{I}} |R_s b_1(y)|^2 \frac{dy\,ds}{s}\}^{1/2}$$

$$\leq (1/\sqrt{|I|}) \{\iint_U |R_s b_1(y)|^2 \frac{dy\,ds}{s}\}^{1/2} = \text{Const}\ \|b_1\|_2/\sqrt{|I|} \leq \text{Const}.$$

Consequently we have $|(R(p),\bar{b})| \leq \text{Const}$. Since b is arbitrary as long as $\|b\|_\infty \leq 1$, we obtain $\|R(p)\|_1 \leq \text{Const}$. Q.E.D.

Lemma 1.9 ([13]). Let $h \in T$ and let E be a subset of an interval I. Then

$$\iint_{\hat{I}-\hat{\Omega}} |h(y,s)|^2 \frac{dy\,ds}{s} \leq \int_{I-E} S(h)(x)^2\,dx,$$

where $\Omega = \{x \in I;\ M\chi_E(x) > 1/2\}$.

Proof. A geometric observation shows that, for any $(y,s) \in \hat{I} - \hat{\Omega}$, $Y \subset \bar{I}$ and $Y \cap \Omega^c \neq \emptyset$, where $Y = [y-s,y+s] (\subset \mathbb{R})$. Let $x_0 \in Y \cap \Omega^c$. Then $|Y \cap E|/|Y| \leq M\chi_E(x_0) \leq 1/2$, and hence $|Y \cap E^c| \geq s$. This shows that

$$\int_{I-E} S(h)(x)^2 dx = \int_{I-E} \{\iint_{\Delta(x)} |h(y,s)|^2 \frac{dy\,ds}{s^2}\}\,dx$$

$$\geq \iint_{\hat{I}-\hat{\Omega}} |h(y,s)|^2 \frac{dy\,ds}{s}.$$ Q.E.D.

We now prove Theorem 1.7. Given $h \in T$, we put $E_k = \{x;\ S(h)(x) > 2^k\}$, $\Omega_k = \{x;\ M\chi_{E_k}(x) > 1/2\}$ $(k = 0, \pm 1, \ldots)$. For each k, let $\{I_j^{(k)}\}_{j=1}^\infty$ be the totality of components of Ω_k and let

$$p_j^{(k)}(y,s) = (2^{-k-1}/|I_j^{(k)}|)\, h(y,s)\, \chi_j^{(k)}(y,s) \qquad (j \geq 1),$$

where $\chi_j^{(k)}$ is the characteristic function of $\hat{I}_j^{(k)} - \hat{\Omega}_j^{(k)}$ $(\Omega_j^{(k)} = I_j^{(k)} \cap \Omega_{k+1})$. Then we obtain the following T-atomic decomposition of h:

$$h(y,s) = \sum_{k=-\infty}^{\infty} \sum_{j=1}^{\infty} 2^{k+1} |I_j^{(k)}| \, p_j^{(k)}(y,s).$$

Each $p_j^{(k)}$ is a T-atom, since $\operatorname{supp}(p_j^{(k)}) \subset \hat{I}_j^{(k)}$ and

$$\iint_{\hat{I}_j^{(k)}} |p_j^{(k)}(y,s)|^2 \frac{dy\,ds}{s} = 2^{-2k-2} |I_j^{(k)}|^{-2} \iint_{\hat{I}_j^{(k)}-\hat{\Omega}_j^{(k)}} |h(y,s)|^2 \frac{dy\,ds}{s}$$

$$\leq 2^{-2k-2} |I_j^{(k)}|^{-2} \int_{I_j^{(k)}-E_{k+1}} S(h)(x)^2\,dx \leq 1/|I_j^{(k)}|$$

by Lemma 1.9. Hence Lemma 1.8 shows that

$$\|R(h)\|_1 \leq \sum_{k=-\infty}^{\infty} \sum_{j=1}^{\infty} 2^{k+1} |I_j^{(k)}| \, \|R(p_j^{(k)})\|_1$$

$$\leq \text{Const} \sum_{k=-\infty}^{\infty} \sum_{j=1}^{\infty} 2^k |I_j^{(k)}| = \text{Const} \sum_{k=-\infty}^{\infty} 2^k |E_k|$$

$$\leq \text{Const} \|S(h)\|_1 = \text{Const} \|h\|_T .$$

This completes the proof of Theorem 1.7. As stated above, Theorem A is deduced from this theorem.

§1.8. The McIntosh expression (Coifman-McIntosh-Meyer [7])

The proof of Theorem A in this section is a version of the method given in [7] for the proof of Theorem B. (See Chapter II.) Here is an interesting expression of $T[a]$.

Lemma 1.10 ([7]). $T[a] = \int_{-\infty}^{\infty} \frac{I}{I + isD} M_a \frac{I}{I + isD} \frac{ds}{s}$ $(a \in L^\infty)$, where I is the identity operator, $D = -i(\partial/\partial x)$ and M_a is the multiplier: $f \to af$.

Proof. Let $a(x) = e^{i\alpha x}$, $f(x) = e^{i\beta x}$ $(\alpha, \beta \in \mathbb{R})$. Then we have

$$T[a]f(x) = (-\pi i)\{ \frac{\alpha+\beta}{\alpha} \operatorname{sign}(\alpha + \beta) - \frac{\beta}{\alpha} \operatorname{sign} \beta \}$$

and

$$\int_{-\infty}^{\infty} \left\{ \frac{I}{I + is\,D} \, M_a \, \frac{I}{I + is\,D} \, f \right\}(x) \, \frac{ds}{s}$$

$$= \int_{-\infty}^{\infty} \left\{ \frac{I}{I + is\,D} \, (af) \right\}(x) \, \frac{1}{1 + is\,\beta} \, \frac{ds}{s}$$

$$= \int_{-\infty}^{\infty} \frac{1}{1 + is(\alpha + \beta)} \, \frac{1}{1 + is\,\beta} \, \frac{ds}{s}$$

$$= \frac{i}{\alpha} \int_{-\infty}^{\infty} \left\{ \frac{1}{1 + is(\alpha + \beta)} - \frac{1}{1 + is\,\beta} \right\} \frac{ds}{s^2}$$

$$= \frac{1}{\alpha} \int_{-\infty}^{\infty} \left\{ \frac{\alpha + \beta}{(1 + is(\alpha + \beta))^2} - \frac{\beta}{(1 + is\,\beta)^2} \right\} \frac{ds}{s}$$

$$= (-\pi i) \left\{ \frac{\alpha + \beta}{\alpha} \operatorname{sign}(\alpha + \beta) - \frac{\beta}{\alpha} \operatorname{sign} \beta \right\}.$$

Hence

$$T[a]f = \int_{-\infty}^{\infty} \left\{ \frac{I}{I + is\,D} \, M_a \, \frac{I}{I + is\,D} \right\} f \, \frac{ds}{s} .$$

Since $\{e^{i\alpha x}\}_{\alpha \in \mathbb{R}}$ is complete in the space of functions f with norm $\int_{-\infty}^{\infty} |f(x)|^2/(1 + x^2)\, dx < \infty$, the required equality holds. (It is necessary to show the convergence of the integral in the right-hand side. This will be given in the proof of Theorem A.) Q.E.D.

Let $P_s = I/(I + s^2D^2)$, $Q_s = sD/(I + s^2D^2)$ $(s > 0)$. In the same manner as in Lemma 1.2, we have

Lemma 1.11 ([7]). Let $a \in \mathrm{BMO}$. Then $|Q_s a(x)|^2 \frac{dxds}{s}$ is a Carleson measure with constant Const $\|a\|_{\mathrm{BMO}}^2$.

Lemma 1.10 shows that

$$(1.14) \quad T[a] = \int_{-\infty}^{\infty} \{P_s M_a P_s - i\, Q_s M_a P_s - i\, P_s M_a Q_s - Q_s M_a Q_s\} \frac{ds}{s}$$

$$= -2i \int_0^{\infty} Q_s M_a P_s \frac{ds}{s} - 2i \int_0^{\infty} P_s M_a Q_s \frac{ds}{s} \quad (= -2i\, L_1 - 2i\, L_2,\ \text{say}).$$

We see that

$$Q_s = 8\, Q_s^3 + s \frac{\partial}{\partial s}\{-Q_s + 2\, P_s Q_s\}, \qquad s \frac{\partial}{\partial s} P_s = -2\, Q_s^2 .$$

Hence the integration by parts shows that

$$L_1 = \int_0^\infty [8\mathcal{Q}_s^3 + s\frac{\partial}{\partial s}\{-\mathcal{Q}_s + 2\mathcal{P}_s\mathcal{Q}_s\}]M_a\mathcal{P}_s\frac{ds}{s}$$

$$= 8\int_0^\infty \mathcal{Q}_s^3 M_a \mathcal{P}_s \frac{ds}{s} - \int_0^\infty \{-\mathcal{Q}_s + 2\mathcal{P}_s\mathcal{Q}_s\} M_a(s\frac{\partial}{\partial s}\mathcal{P}_s)\frac{ds}{s}$$

$$= 8\int_0^\infty \mathcal{Q}_s^3 M_a \mathcal{P}_s \frac{ds}{s} + 2\int_0^\infty\{-\mathcal{Q}_s + 2\mathcal{Q}_s\mathcal{P}_s\}M_a\mathcal{Q}_s^2\frac{ds}{s}$$

$$(= 8\,L_{11} + 2\,L_{12},\quad \text{say}).$$

Since $\|\mathcal{P}_s\|_{2,2} \leqq$ Const, $\|\mathcal{Q}_s\|_{2,2} \leqq$ Const and $\int_0^\infty \|\mathcal{Q}_s f\|_2^2 \frac{ds}{s} =$ Const $\|f\|_2^2$, Shwartz's inequality shows that $\|L_{12}\|_{2,2} \leqq$ Const $\|a\|_\infty$. We have, for $f,g \in L^2$,

$$|(g, L_{11}f)| = |\int_0^\infty (\mathcal{Q}_s^2 g, \overline{\mathcal{Q}_s M_a \mathcal{P}_s f})\frac{ds}{s}|$$

$$\leqq \{\int_0^\infty \|\mathcal{Q}_s^2 g\|_2^2 \frac{ds}{s}\}^{1/2}\{\int_0^\infty \|\mathcal{Q}_s M_a \mathcal{P}_s f\|_2^2\frac{ds}{s}\}^{1/2}$$

$$= \text{Const}\,\|g\|_2\{\int_0^\infty \|\mathcal{Q}_s M_a \mathcal{P}_s f\|_2^2 \frac{ds}{s}\}^{1/2}.$$

We see that

$$\{\mathcal{Q}_s M_a \mathcal{P}_s\}f = (\mathcal{Q}_s a)(\mathcal{P}_s f) + \mathcal{P}_s\{(\mathcal{P}_s a)(\mathcal{Q}_s f)\} - \mathcal{Q}_s\{(\mathcal{Q}_s a)(\mathcal{Q}_s f)\}.$$

(To see this, use $a(x) = e^{i\alpha x}$, $f(x) = e^{i\beta x}$ $(\alpha, \beta \in R)$.) Hence we have

$$\|\mathcal{Q}_s M_a \mathcal{P}_s f\|_2^2 \leqq \text{Const}\,\{\|a\|_\infty^2 \|\mathcal{Q}_s f\|_2^2 + \|(\mathcal{Q}_s a)(\mathcal{Q}_s f)\|_2^2\}.$$

Lemma 1.11 shows that

$$\int_0^\infty \|\mathcal{Q}_s M_a \mathcal{P}_s f\|_2^2 \frac{ds}{s} \leqq \text{Const}\,\|a\|_\infty^2 \int_0^\infty \|\mathcal{Q}_s f\|_2^2\frac{ds}{s}$$

$$+ \text{Const} \iint_U |\mathcal{Q}_s a(x)\,\mathcal{P}_s f(x)|^2 \frac{dx\,ds}{s}$$

$$\leqq \text{Const}\,\|a\|_\infty^2\|f\|_2^2 + \text{Const}\iint_U |\mathcal{Q}_s a(x)\,P_s * f(x)|^2\frac{dx\,ds}{s}$$

$$\leqq \text{Const}\,\{\|a\|_\infty^2 + \|a\|_{BMO}^2\}\,\|f\|_2^2 \leqq \text{Const}\,\|a\|_\infty^2\,\|f\|_2^2,$$

which gives $\|L_{11}\|_{2,2} \leqq$ Const $\|a\|_\infty$. Thus $\|L_1\|_{2,2} \leqq$ Const $\|a\|_\infty$. Since L_2 is the dual operator of $\overline{L_1}$, we have $\|L_2\|_{2,2} = \|L_1\|_{2,2}$. Consequently, (1.14) gives (1.4).

§1.9. Almost orthogonality (David-Journé [19])

David-Journé [19] showed the so-called T1 theorem. David-Journé-Semmes [20] showed the so-called Tb theorem (cf. McIntosh-Meyer [40]). These theorems give immediately Theorem A. Given $\delta > 0$, we use C_δ for various constants depending only on δ; the value of C_δ differs in general from one occurrence to another. For $0 < \delta \leqq 1$, we say that a kernel $K(x,y)$ $(x \neq y;\ x,y \in \mathbb{R})$ is a δ-standard kernel if there exists $B > 0$ such that

$$|K(x,y)| \leq B/|x-y|,$$

$$|K(x,y) - K(x',y)| \leq B|x-x'|^{\delta} / |x-y|^{1+\delta} \qquad (|x-x'| \leq |x-y|/2),$$

$$|K(x,y) - K(x,y')| \leq B|y-y'|^{\delta} / |x-y|^{1+\delta} \qquad (|y-y'| \leq |x-y|/2).$$

We denote by $\omega_\delta(K)$ the minimum of constants B satisfying the above three inequalities. For the sake of simplicity, we assume that

(1.15) $K(x,y)$ is anti-symmetric, i.e., $K(x,y) = -K(y,x)$,

(1.16) $$K1(x) = \lim_{\varepsilon \to 0} \int_{\varepsilon < |x-y| < 1} K(x,y)\, dy + \lim_{\varepsilon \to 0} \int_{1 < |x-y| < 1/\varepsilon} K(x,y) dy \quad \text{exists a.e.}$$

We write simply K the operator defined by the kernel $K(x,y)$. The following theorem is a special case of the T1 theorem; we rewrite the T1 theorem so that only the proof of Theorem A can be given.

Theorem 1.12 ([19]). Let $K(x,y)$ be a δ-standard kernel satisfying (1.15) and (1.16). Then $\|K\|_{2,2} \leq C_\delta \{\|K1\|_{BMO} + \omega_\delta(K)\}$.

Integration by parts shows that $T[a]1 = \pi H a$. We see that $\|Ha\|_{BMO} \leq \text{Const } \|a\|_\infty$ and $\omega_1(T[a]) \leq \text{Const } \|a\|_\infty$. (See Lemma 2.5.) Hence this theorem immediately yields Theorem A.

Lemma 1.13 ([19]). For $b \in BMO$, there exists an anti-symmetric 1-standard kernel $L(x,y)$ such that $L1 = b$, $\|L\|_{2,2} \leq \text{Const } \|b\|_{BMO}$ and $\omega_1(L) \leq \text{Const } \|b\|_{BMO}$.

Proof. Let L be an operator defined by

$$Lf = 2\int_0^\infty \mathcal{Q}_s\{(\mathcal{Q}_s b)(\mathcal{P}_s f)\} \frac{ds}{s} + 2\int_0^\infty \mathcal{P}_s\{(\mathcal{Q}_s b)(\mathcal{Q}_s f)\} \frac{ds}{s}.$$

Then its kernel $L(x,y)$ is given by

$$L(x,y) = \text{Const } \Big\{\iint_U v_s(x-t)(v_s*b)(t)u_s(t-y) \frac{dt\, ds}{s} + \iint_U u_s(x-t)(v_s*b)(t)\, v_s(t-y) \frac{dt\, ds}{s}\Big\},$$

where $u_s(x) = (1/s)e^{-|x|/s}$, $v_s(x) = \text{sign}(x/s)u_s(x)$ $(x \in \mathbb{R},\ s > 0)$ and $U = \{(t,s);\ t \in \mathbb{R},\ s > 0\}$. Then $L(x,y)$ is anti-symmetric and

$$L1 = 2\int_0^\infty \mathcal{Q}_s\mathcal{Q}_s b \frac{ds}{s} = b.$$

Since $|v_s*b(t)|^2\, dtds/s$ is a Carleson measure with constant $\text{Const } \|b\|^2_{BMO}$, we have $\|L\|_{2,2} \leq \text{Const } \|b\|_{BMO}$.

It remains to prove $\omega_1(L) \leq \text{Const } \|b\|_{BMO}$. In the same manner as in

Lemma 1.2, we have $\|v_s * b\|_\infty \leqq \text{Const}\,\|b\|_{BMO}$. Since $|v_s(x)| \leqq u_s(x)$,

$$|L(x,y)| \leqq \text{Const}\,\|b\|_{BMO} \iint_U u_s(x-t)u_s(t-y)\,\frac{dt\,ds}{s}$$

$$= \text{Const}\,\|b\|_{BMO} \int_{-\infty}^{\infty} \frac{dt}{(|x-t| + |t-y|)^2} \int_0^\infty \frac{1}{s^3}\, e^{-1/s}\, ds$$

$$\leqq \text{Const}\,\|b\|_{BMO}/|x-y| .$$

Since $|u_s'(x)| \leqq u_s(x)/s$, $|v_s'(x)| \leqq u_s(x)/s$ $(x \neq 0)$, we have

$$\left|\frac{\partial}{\partial x} L(x,y)\right| \leqq \text{Const}\,\|b\|_{BMO} \iint_U u_s(x-t)u_s(y-t)\,\frac{dt\,ds}{s^2}$$

$$= \text{Const}\,\|b\|_{BMO} \int_{-\infty}^{\infty} \frac{dt}{(|x-t| + |t-y|)^3} \int_0^\infty \frac{1}{s^4}\, e^{-1/s}\, ds$$

$$\leqq \text{Const}\,\|b\|_{BMO}/|x-y|^2.$$

Thus $\omega_1(L) \leqq \text{Const}\|b\|_{BMO}$. Q.E.D.

Here is the main tool for the proof of Theorem 1.12. We say that an operator L from L^2 to itself is anti-self adjoint if $(Lf,g) = -(f,\overline{L}g)$ for any $f, g \in L^2$.

Lemma 1.14 (Cotlar's Lemma [16]). Let $\rho(t)$ be a function from $[0,\infty)$ to itself and let $\{L_k\}_{k=-N}^{N}$ be anti-self adjoint operators from L^2 to itself such that $\|L_jL_k\|_{2,2} \leqq \rho(|j-k|)$ $(j,k = 0, \pm 1, \ldots, \pm N)$. Then $\|\Sigma_{k=-N}^{N} L_k\|_{2,2} \leqq \text{Const}\ \Sigma_{k=0}^{2N} \sqrt{\rho(k)}$.

Proof. The following proof is due to Fefferman [25]. Let $L = \Sigma_{k=-N}^{N} L_k$. Then, for any $M \geqq 1$, $\|L^{2M}\|_{2,2}^{1/2M} = \|L\|_{2,2}$. We have

$$L^{2M} = \sum_{-N \leqq k_1,\ldots,k_{2M} \leqq N} L_{k_1}L_{k_2} \cdots L_{k_{2M}} .$$

Since

$$\|L_{k_1}L_{k_2} \cdots L_{k_{2M}}\|_{2,2} \leqq \|L_{k_1}L_{k_2}\|_{2,2} \cdots \|L_{k_{2M-1}}L_{k_{2M}}\|_{2,2}$$

$$\leqq \rho(|k_1-k_2|) \cdots \rho(|k_{2M-1} - k_{2M}|)$$

and

$$\|L_{k_1}\cdots L_{k_{2M}}\|_{2,2} \leqq \|L_{k_1}\|_{2,2}\,\|L_{k_2}L_{k_3}\|_{2,2} \cdots \|L_{k_{2M-2}}L_{k_{2M-1}}\|_{2,2}\,\|L_{k_{2M}}\|_{2,2}$$

$$\leqq \sqrt{\rho(0)}\ \rho(|k_2-k_3|) \cdots \rho(|k_{2M-2}-k_{2M-1}|)\ \sqrt{\rho(0)} ,$$

we have

$$\|L_{k_1}\cdots L_{k_{2M}}\|_{2,2}^2 \leq \rho(0)\,\rho(|k_1-k_2|)\,\rho(|k_2-k_3|)\ \ldots\ \rho(|k_{2M-1}-k_{2M}|).$$

Thus

$$\|L\|_{2,2} = \|L^{2M}\|_{2,2}^{1/2M}$$

$$\leq \{\sqrt{\rho(0)} \sum_{-N \leq k_1,\ldots,k_{2M} \leq N} \sqrt{\rho(|k_1-k_2|)\rho(|k_2-k_3|)\ \ldots\rho(|k_{2M-1}-k_{2M}|)}\ \}^{1/2M}$$

$$\leq \{\sqrt{\rho(0)} \sum_{-N \leq k_1,\ldots,k_{2M-1} \leq N} \sqrt{\rho(|k_1-k_2|)\ldots\rho(|k_{2M-2}-k_{2M-1}|)} \times (2 \sum_{j=0}^{2N} \sqrt{\rho(j)})\ \}^{1/2M}$$

$$\leq \ldots \leq \{\sqrt{\rho(0)}\,(2N+1)(2 \sum_{j=0}^{2N} \sqrt{\rho(j)})^{2M}\}^{1/2M}$$

$$\leq \rho(0)^{1/4M}\,(2N+1)^{1/2M}\ 2 \sum_{j=0}^{2N} \sqrt{\rho(j)}\ .$$

Letting M tend to infinity, we obtain the required inequality. Q.E.D.

We now give the proof of Theorem 1.12. We may assume that $K(x,y)$ is real-valued. Since $K1 \in BMO$, we can define, by Lemma 1.13, an anti-symmetric kernel $L(x,y)$ so that $K1 = L1$, $\|L\|_{2,2} \leq \text{Const}\,\|K1\|_{BMO}$ and $\omega_\delta(L) \leq \omega_1(L) \leq \text{Const}\,\|K1\|_{BMO}$. Consider the kernel $K(x,y) - L(x,y)$. Then this is anti-symmetric and satisfies $(K-L)1 = 0$,

$$\omega_\delta(K - L) \leq \text{Const}\,\{\|K1\|_{BMO} + \omega_\delta(K)\}\ .$$

Hence from the beginning, we assume $K1 = 0$, and show $\|K\|_{2,2} \leq C_\delta \omega_\delta(K)$. To do this, we may assume that $\omega_\delta(K) = 1$. Choosing $h \in C_0^\infty$ so that

$$0 \leq h(x) \leq 1,\quad h(x) = h(-x),\quad \text{supp}(h) \subset [-1,1],\quad \|h\|_1 = 1,$$

we put

$$K_k(x,y) = \int_{-\infty}^{\infty}\int_{-\infty}^{\infty} K(x-s,y-t)\{h_k(s)h_k(t) - h_{k+1}(s)h_{k+1}(t)\}\,ds\,dt \quad (k = 0,\pm 1,\ldots),$$

where $h_k(x) = 2^{-k}h(2^{-k}x)$. Then K_k is anti-self adjoint, $K_k 1 = 0$ and $K = \lim_{k\to\infty} \sum_{k=-N}^{N} K_k$. We show that

(1.17) $$|K_k(x,y)| \leq C_\delta\ 2^{-k}/\{1 + |x-y|2^{-k}\}^{1+\delta},$$

$$(1.18)\qquad \left|\frac{\partial}{\partial x} K_k(x,y)\right| \leqq C_\delta\ 2^{-2k}/\{1+|x-y|2^{-k}\}^{1+\delta}.$$

If $|x-y| \geqq 4\cdot 2^k$, then we have

$$\begin{aligned}
&|K_k(x,y)| \\
&= \left|\int_{-\infty}^{\infty}\int_{-\infty}^{\infty}\{K(x-s,y-t)-K(x,y)\}\{h_k(s)h_k(t)-h_{k+1}(s)h_{k+1}(t)\}\ ds\,dt\right| \\
&\leqq \int_{-\infty}^{\infty}\int_{-\infty}^{\infty} C_\delta\ \frac{|s|^\delta+|t|^\delta}{|x-y|^{1+\delta}}\ \{h_k(s)h_k(t)+h_{k+1}(s)h_{k+1}(t)\}\ ds\,dt \\
&\leqq C_\delta\ 2^{\delta k}/|x-y|^{1+\delta} \leqq C_\delta\ 2^{-k}/\{1+|x-y|2^{-k}\}^{1+\delta}.
\end{aligned}$$

Let I, J be two intervals in an interval L. Since $K(x,y)$ is anti-symmetric, we have $\int_{I\cap J}\int_{I\cap J} K(s,t)\,ds\,dt = 0$, and hence

$$(1.19)\qquad \left|\int_I \{\int_J K(s,t)ds\}dt\right| = \Big|\int_{I-(I\cap J)} \{\int_J K(s,t)ds\}\ dt$$

$$+ \int_{I\cap J} \{\int_{J-(I\cap J)} K(s,t)ds\}dt\ \Big| \leqq 2\int_0^{|L|}\int_0^{|L|} \frac{ds\ dt}{s+t} \leqq \text{Const}\ |L|.$$

Integration by parts and (1.19) show that, if $|x-y| < 4\cdot 2^k$, then

$$\begin{aligned}
&|K_k(x,y)| \\
&= \left|\int_{-\infty}^{\infty}\int_{-\infty}^{\infty}\{\int_0^u\int_0^v K(x-s,y-t)ds\,dt\}\ \{h_k'(u)h_k'(v)-h_{k+1}'(u)h_{k+1}'(v)\}\ du\,dv\right| \\
&\leqq \int_{-\infty}^{\infty}\int_{-\infty}^{\infty} (\text{Const}\ 2^k)\ \{|h_k'(u)h_k'(v)| + |h_{k+1}'(u)h_{k+1}'(v)|\}\ du\,dv \\
&\leqq \text{Const}\ 2^{-k} \leqq C_\delta\ 2^{-k}/\{1+|x-y|2^{-k}\}^{1+\delta}.
\end{aligned}$$

Thus (1.17) holds. If $|x-y| \geqq 4\cdot 2^k$, then

$$\begin{aligned}
&\left|\frac{\partial}{\partial x} K_k(x,y)\right| \\
&= \left|\int_{-\infty}^{\infty}\int_{-\infty}^{\infty}\{K(x-s,y-t)-K(x,y)\}\ \{h_k'(s)h_k(t)-h_{k+1}'(s)h_{k+1}(t)\}\ ds\,dt\right| \\
&\leqq \int_{-\infty}^{\infty}\int_{-\infty}^{\infty} C_\delta\ \frac{|s|^\delta+|t|^\delta}{|x-y|^{1+\delta}}\ \{|h_k'(s)|h_k(t)+|h_{k+1}'(s)|h_{k+1}(t)\}\ ds\,dt \\
&\leqq C_\delta\ 2^{(\delta-1)k} \leqq C_\delta\ 2^{-2k}/\{1+|x-y|2^{-k}\}^{1+\delta}.
\end{aligned}$$

If $|x-y| < 4\cdot 2^k$, then

$$\left|\frac{\partial}{\partial x} K_k(x,y)\right|$$

$$= \left|\int_{-\infty}^{\infty}\int_{-\infty}^{\infty}\{\int_0^u \int_0^v K(x-s,y-t)dsdt\}\{h_k''(u)h_k'(v) - h_{k+1}''(u)h_{k+1}'(v)\}\, du\, dv\right|$$

$$\leqq \text{Const } 2^{-2k} \leqq C_\delta\, 2^{-2k}/\{1 + |x-y|2^{-k}\}^{1+\delta} .$$

Thus (1.18) holds.

We now show that, for $k \geqq \ell$,

(1.20) $\quad |(K_k K_\ell)(x,y)| \leqq C_\delta\, \rho_{k,\ell}(x-y),$

where

$$\rho_{k,\ell}(s) = 2^{-(k+\ell)/2}(1 + |s|2^{-\ell})^{-(1/2)-\delta}$$
$$+2^{(\ell\delta/2)}\{2^{-2k}|s|^{1-(\delta/2)} + 2^{-k}|s|^{-\delta/2}\}(1 + |s|\,2^{-k})^{-1-\delta} .$$

Let I be the interval of midpoint y and of length $|x-y|/2$. By (1.17), we have

$$\left|\int_{I^c} K_k(x,s)K\,(s,y)ds\right| \leqq C_\delta \int_{I^c} \frac{2^{-k}}{(1 + |x-s|2^{-k})^{1+\delta}}\,\frac{2^{-\ell}}{(1 + |s-y|2^{-\ell})^{1+\delta}}\,ds$$

$$\leqq C_\delta \frac{2^{-\ell/2}}{(1 + |x-y|2^{-\ell})^{(1/2)+\delta}} \int_{-\infty}^{\infty} \frac{2^{-k}}{(1 + |x-s|2^{-k})^{1+\delta}|x-s|^{1/2}}\,ds \leqq C_\delta\rho_{k,\ell}\,(x-y).$$

By $K_\ell 1(y) = 0$, we have

$$\left|\int_I K_k(x,s)K_\ell(s,y)ds\right| \leqq \left|\int_I \{K_k(x,s)-K_k(x,y)\}\, K_\ell(s,y)ds\right|$$

$$+ \left|K_k(x,y)\int_I K_\ell(s,y)ds\right| = \left|\int_I \{K_k(x,s) - K_k(x,y)\}\, K_\ell(s,y)ds\right|$$

$$+ \left|K_k(x,y)\int_{I^c} K_\ell(s,y)ds\right| \quad (= L_1(x,y) + L_2(x,y), \text{ say}).$$

By (1.17) and (1.18), we have

$$L_1(x,y) \leqq C_\delta \int_I \frac{2^{-2k}|s-y|}{(1 + |x-y|2^{-k})^{1+\delta}}\,\frac{2^{-\ell}}{(1 + |s-y|2^{-\ell})^{1+\delta}}\,ds$$

$$\leqq C_\delta \frac{2^{-2k+(\ell\delta/2)}}{(1 + |x-y|2^{-k})^{1+\delta}} \int_I |s-y|^{-\delta/2}\,ds \leqq C_\delta\, \rho_{k,\ell}(x-y)$$

and

$$L_2(x,y) \leqq C_\delta \frac{2^{-k}}{(1+|x-y|2^{-k})^{1+\delta}} \int_{I^c} \frac{2^{-\ell}}{(1+|s-y|2^{-\ell})^{1+\delta}} ds$$

$$\leqq C_\delta \frac{2^{-k+(\ell\delta/2)}}{(1+|x-y|2^{-k})^{1+\delta}} \int_{I^c} |s-y|^{-1-(\delta/2)} ds \leqq C_\delta \rho_{k,\ell}(x-y).$$

Thus (1.20) holds. Since $\|\rho_{k,\ell}\|_1 \leqq C_\delta 2^{-(k-\ell)/2}$, we have $\|K_k K_\ell\|_{2,2} \leqq C_\delta 2^{-(k-\ell)/2}$. Hence Lemma 1.14 gives $\|\Sigma_{k=-N}^{N} K_k\|_{2,2} \leqq C_\delta$ $(N \geqq 1)$. Letting N tend to infinity, we obtain $\|K\|_{2,2} \leqq C_\delta$. This completes the proof of Theorem 1.12.

§1.10. Interpolation (Lemarie [36])

In this section, we give a proof of Theorem 1.12 (in the case of $K1 = 0$) by interpolation, which was given by Lemarie. For $\alpha \in \mathbb{R}$ with $|\alpha| < 1$, let E_α denote the Banach space of distributions obtained from the completion of C_0^∞ with respect to the norm $|||f|||_\alpha = \{\int_{-\infty}^{\infty} |\xi|^\alpha |\hat{f}(\xi)|^2 d\xi\}^{1/2}$, where $\hat{f}$ is the Fourier transform of f (in the sense of distributions). For $0 < \alpha < 1$, let $[E_\alpha, E_{-\alpha}]$ denote the Banach space of distributions f in $E_\alpha \oplus E_{-\alpha}$ with norm

$$\|f\|_{Int,\alpha} = \{\int_0^\infty B(s,f)^2 \frac{ds}{s^2}\}^{1/2} < \infty ,$$

where

$$B(s,f) = \inf\{(|||g|||_\alpha^2 + s^2 |||h|||_{-\alpha}^2)^{1/2};\ f = g+h,\ g \in E_\alpha,\ h \in E_{-\alpha}\}.$$

Lemarie showed the following two facts.

Lemma 1.15 ([36]). Let $K(x,y)$ be a δ-standard kernel satisfying (1.15), (1.16) and $K1 = 0$. Then, for any $0 < \alpha < \min\{1, 2\delta\}$, $|||K|||_{\alpha,\alpha} \leqq C_{\alpha,\delta}\, \omega_\delta(K)$, where $|||K|||_{\alpha,\alpha}$ is the norm of K as an operator from E_α to itself and $C_{\alpha,\delta}$ is a constant depending only on α and δ.

Proof. Throughout the proof, we use $C_{\alpha,\delta}$ for various constants depending only on α and δ. We may assume that $\omega_\delta(K) = 1$. Note that

$$|||f|||_\alpha^2 = C_\alpha \int_{-\infty}^{\infty}\int_{-\infty}^{\infty} \frac{|f(x)-f(y)|^2}{|x-y|^{1+\alpha}} dx\, dy \qquad (f \in E_\alpha).$$

Given $x, y \in \mathbb{R}$, we write by I the interval of midpoint $(x+y)/2$ and of length $2|x-y|$. By $K1 = 0$, we have

$$|Kf(x) - Kf(y)| = |\int_{-\infty}^{\infty}\{K(x,s) - K(y,s)\} f(s)ds|$$

$$= |\int_{-\infty}^{\infty}K(x,s)(f(s)-f(x))ds - \int_{-\infty}^{\infty} K(y,s)(f(s)-f(y))ds|$$

$$= |\int_I K(x,s)(f(s)-f(x))ds - \int_I K(y,s)(f(s) - f(y))ds$$

$$+ \int_{I^c} \{K(x,s) - K(y,s)\} (f(s) - f(x))ds - (f(x) - f(y)) \int_{I^c} K(y,s)ds|$$

$$\leqq \int_I |f(s) - f(x)|/|s-x| ds + \int_I |f(s) - f(y)|/|s-y| ds$$

$$+ C_\delta \int_{I^c} |f(s)-f(x)| |x-y|^\delta/|s-x|^{1+\delta} ds + |f(x)-f(y)| |\int_{I^c} K(y,s)ds|$$

$$(= L_1(x,y) + L_2(x,y) + C_\delta L_3(x,y) + L_4(x,y), \text{ say}).$$

Hence

$$|||Kf|||_\alpha^2 = C_\alpha \int_{-\infty}^{\infty} \int_{-\infty}^{\infty} \frac{|Kf(x)-Kf(y)|^2}{|x-y|^{1+\alpha}} dx\, dy$$

$$\leqq C_{\alpha,\delta} \sum_{k=1}^{4} \int_{-\infty}^{\infty} \int_{-\infty}^{\infty} \frac{L_k(x,y)^2}{|x-y|^{1+\alpha}} dx\, dy \quad (= C_{\alpha,\delta} \sum_{k=1}^{4} L_k', \text{ say}).$$

Choosing $0 < \beta < 1/2$ so that $\alpha + 2\beta > 1$, we have

$$L_1(x,y)^2 \leqq C_\alpha \int_{|s-x| < 2|x-y|} \frac{|f(s)-f(x)|^2}{|s-x|^{2(1-\beta)}} ds \quad |x-y|^{1-2\beta} ,$$

and hence

$$L_1' \leqq C_\alpha \int_{-\infty}^{\infty} \int_{-\infty}^{\infty} \frac{|f(s)-f(x)|^2}{|s-x|^{2(1-\beta)}} \{ \int_{2|x-y| > |s-x|} \frac{|x-y|^{1-2\beta}}{|x-y|^{1+\alpha}} dy \} ds\, dx$$

$$= C_\alpha |||f|||_\alpha^2 .$$

In the same manner, $L_2' \leqq C_\alpha |||f|||_\alpha^2$. Choosing $0 < \gamma < 1$ so that $2(1 + \delta) \gamma > 1$ and $(\alpha - 2\delta) + 2(1 + \delta) \gamma < 1$, we have

$$L_3(x,y)^2 \leqq C_{\alpha,\delta} \int_{|s-x| > |x-y|/2} \frac{|f(s)-f(x)|^2}{|s-x|^{2(1+\delta)(1-\gamma)}} |x-y|^{2\delta-2(1+\delta)\gamma + 1},$$

and hence

$$L_3' \leqq C_{\alpha,\delta} \int_{-\infty}^{\infty} \int_{-\infty}^{\infty} \frac{|f(s)-f(x)|^2}{|s-x|^{2(1+\delta)(1-\gamma)}} \{\int_{|x-y|/2< |s-x|} |x-y|^{2\delta-2(1+\delta)\gamma-\alpha} dy\} ds dx$$

$$= C_{\alpha,\delta} |||f|||_\alpha^2 .$$

Since $|\int_{I^c} K(y,s)ds| \leqq |\int_{|y-s|>|x-y|/2} K(y,s)ds| + \text{Const}$, we have

$$L_4' \leqq \text{Const}\ |||f|||_\alpha \{ \sup_{y\in\mathbb{R},\varepsilon>0} |\int_{|y-s|>\varepsilon} K(y,s)ds| + 1\} .$$

Given $y_0 \in \mathbb{R}$, $\varepsilon > 0$, we have, with $J = (y_0 - \varepsilon, y_0 + \varepsilon)$ and $J^* = (y_0 - 2\varepsilon, y_0 + 2\varepsilon)$,

$$|\int_{J^c} K(y_0,s)ds| = |\int_{J^c} K(y_0,s)ds + \frac{1}{|J|} \int_J \int_J K(y,s)ds\, dy|$$

$$= \left| \frac{1}{|J|} \int_J \{\int_{J^c} (K(y_0,s) - K(y,s))ds\}\, dy\right| = \frac{1}{|J|} |\int_J \{\int_{J^*-J} + \int_{J^{*c}} \}|$$

$$\leqq (\text{Const}/|J|) \int_J \{\int_{J^*-J} \frac{ds}{|y-s|} \}dy$$

$$+ (C_\delta/|J|) \int_J \{ \int_{J^{*c}} \frac{|y_0-y|^\delta}{|y-s|^{1+\delta}} ds\}\ dy \leqq C_\delta .$$

Hence $L_4' \leqq C_{\alpha,\delta}\ |||f|||_\alpha^2$. Consequently we have $|||Kf|||_\alpha^2 \leqq C_{\alpha,\delta}\ |||f|||_\alpha^2$, which gives the required inequality. Q.E.D.

Lemma 1.16 ([36]). For any $0 < \alpha < 1$, $[E_\alpha, E_{-\alpha}] = L^2$. More precisely, $(1/C_\alpha)\|f\|_2 \leqq \|f\|_{Int,\alpha} \leqq C_\alpha \|f\|_2$ $(f \in L^2)$.

Proof. Given $f \in L^2$, $s > 0$, we put

$$g_s = \frac{2}{\sqrt{\pi}} \int_{s^{-1/\alpha}}^\infty \frac{D}{I + t^2D^2} f\, dt, \quad h_s = \frac{2}{\sqrt{\pi}} \int_0^{s^{-1/\alpha}} \frac{D}{I + t^2D^2} f\ dt.$$

Then $f = g_s + h_s$ and

$$\|f\|_{Int,\alpha}^2 = \int_0^\infty B(s,f)^2 \frac{ds}{s^2} \leqq \int_0^\infty \{|||g_s|||_\alpha^2 + s^2\ |||h_s|||_{-\alpha}^2 \} \frac{ds}{s^2}$$

$$= C_\alpha \int_0^\infty [\int_{-\infty}^\infty \{(\int_{s^{-1/\alpha}}^\infty \frac{|\xi|}{1 + t^2\xi^2} dt)^2 \frac{|\xi|^\alpha}{s^2}$$

$$+ (\int_0^{s^{-1/\alpha}} \frac{|\xi|}{1 + t^2\xi^2} dt)^2 \frac{1}{|\xi|^\alpha} \} |\hat{f}(\xi)|^2 d\xi] ds$$

$$\leqq C_\alpha \int_0^\infty [\int_{-\infty}^\infty \{(\int_{s^{-1/\alpha}}^\infty \frac{|\xi|}{1 + t^2\xi^2} dt) \frac{|\xi|^\alpha}{s^2}$$

$$+ (\int_0^{s^{-1/\alpha}} \frac{|\xi|}{1 + t^2\xi^2} dt) \frac{1}{|\xi|^\alpha} \} |\hat{f}(\xi)|^2 d\xi] ds$$

$$= C_\alpha \int_{-\infty}^\infty |\hat{f}(\xi)|^2 \{\int_0^\infty (\int_{t^{-\alpha}}^\infty \frac{ds}{s^2}) \frac{|\xi|^{1+\alpha}}{1 + t^2\xi^2} dt + \int_0^\infty (\int_0^{t^{-\alpha}} ds) \frac{|\xi|^{1-\alpha}}{1+t^2|\xi|^2} dt\} d\xi$$

$$= C_\alpha \int_{-\infty}^\infty |\hat{f}(\xi)|^2 d\xi = C_\alpha \|f\|_2^2 .$$

Let $f \in [E_\alpha, E_{-\alpha}]$. For each $s > 0$, we choose $g_s \in E_\alpha$, $h_s \in E_{-\alpha}$ so that $f = g_s + h_s$ and $|||g_s|||_\alpha^2 + s^2 |||h_s|||_{-\alpha}^2 \leq 2B(s,f)^2$. Then

$$\|f\|_2^2 = \text{Const} \int_0^\infty \| \frac{tD}{I + (tD)^2} f\|_2^2 \frac{dt}{t}$$

$$\leq \text{Const} \int_0^\infty \{ \| \frac{tD}{I + (tD)^2} g_{t^{-\alpha}} \|_2^2 + \| \frac{tD}{I + (tD)^2} h_{t^{-\alpha}} \|_2^2 \} \frac{dt}{t}$$

$$= \text{Const} \int_0^\infty [\int_{-\infty}^\infty \frac{(t\xi)^2}{(1 + (t\xi)^2)^2} \{ |\hat{g}_{t^{-\alpha}}(\xi)|^2 + |\hat{h}_{t^{-\alpha}}(\xi)|^2 \} d\xi] \frac{dt}{t}$$

$$\leq \text{Const} \int_0^\infty [\int_{-\infty}^\infty \{ t^\alpha |\xi|^\alpha |\hat{g}_{t^{-\alpha}}(\xi)|^2 + t^{-\alpha} |\xi|^{-\alpha} |\hat{h}_{t^{-\alpha}}(\xi)|^2 \} d\xi] \frac{dt}{t}$$

$$= C_\alpha \int_0^\infty (|||g_{t^{-\alpha}}|||_\alpha^2 + t^{-2\alpha} |||h_{t^{-\alpha}}|||_{-\alpha}^2) \frac{dt}{t^{1-\alpha}}$$

$$= C_\alpha \int_0^\infty (|||g_s|||_\alpha^2 + s^2 |||h_s|||_{-\alpha}^2) \frac{ds}{s^2} \leq C_\alpha \int_0^\infty B(s,f)^2 \frac{ds}{s^2} = C_\alpha \|f\|_{\text{Int},\alpha}^2 .$$

Q.E.D.

Theorem 1.12 is deduced as follows. We may assume that $K1 = 0$ and $\omega_\delta(K) = 1$. We use Lemmas 1.15 and 1.16 with $\alpha = \delta/2$. Let $f \in L^2$. Then, for each $s > 0$, we can choose $g_s \in E_\alpha$, $h_s \in E_{-\alpha}$ so that $f = g_s + h_s$, $|||g_s|||_\alpha^2 + |||h_s|||_{-\alpha}^2 \leq 2\, B(s,f)^2$, by Lemma 1.15. Thus Lemmas 1.15 and 1.16 show that

$$\|Kf\|_2^2 \leq C_\delta \|Kf\|_{\text{Int},\alpha}^2 = C_\delta \int_0^\infty B(s,Kf)^2 \frac{ds}{s^2}$$

$$\leq C_\delta \int_0^\infty (|||Kg_s|||_\alpha^2 + s^2 |||Kh_s|||_{-\alpha}^2) \frac{ds}{s^2}$$

$$\leq C_\delta \int_0^\infty (|||g_s|||_\alpha^2 + s^2 |||h_s|||_{-\alpha}^2) \frac{ds}{s^2} \leq C_\delta \int_0^\infty B(s,f)^2 \frac{ds}{s^2}$$

$$= C_\delta \| f\|_{\text{Int},\alpha}^2 \leq C_\delta \|f\|_2^2.$$

§1.11 Successive compositions of kernels

Meyer [41] also gave a proof of Theorem 1.12 from the point of view of composition of kernels. Using his method, we show the following lemma which also yields Theorem 1.12.

Lemma 1.17. Let $K(x,y)$ be a δ-standard kernel satisfying (1.15), (1.16), $K1 = 0$ and

$$\sup_{x,y \in \mathbb{R}} |K(x,y)| (1 + |x-y|)^{1+\delta} < \infty. \tag{1.21}$$

We define kernels $\{K^{(n)}(x,y)\}_{n=1}^{\infty}$ by

$$K^{(1)}(x,y) = K(x,y), \quad K^{(n)}(x,y) = \int_{-\infty}^{\infty} K(x,s)K^{(n-1)}(s,y)ds \qquad (n \geqq 2),$$

and define

$$\beta(K^{(n)}) = \sup \{ \frac{1}{|I|} \ |\int_I \ K^{(n)} \chi_I(x)dx| \ ; \ I \ \text{interval}\} \qquad (n \geqq 1).$$

Then, for any $0 < \varepsilon < \delta$,

$$\beta(K^{(n)}) + \omega_{\delta-\varepsilon}(K^{(n)}) \leqq C_{\delta,\varepsilon}^{n} \ \omega_\delta(K)^n ,$$

where $C_{\delta,\varepsilon}$ is a constant depending only on δ, ε.

Postponing the proof later, we now deduce Theorem 1.12 (in the case of $K1 = 0$) from this lemma. (This lemma plays the role of Cotlar's lemma.) Without loss of generality, we may assume that $K(x,y)$ is real-valued, $\omega_\delta(K) = 1$ and $K1 = 0$. Using $K_k(x,y)$ in §1.9, if necessary, we may assume that $K(x,y)$ satisfies (1.21). We put

$$(1.22) \qquad \sigma(K) = \sup \{ \frac{1}{|I|} \ \sigma(I,K,f); \ f \in L^{\infty}_{real,1} \ \ I \ \text{interval}\},$$

where, in general,

$$(1.23) \qquad L^{\infty}_{real,\beta} = \{f \in L^{\infty}; \ \|f\|_\infty \leqq \beta , \ \ f \ \text{real-valued} \ \} \quad (\beta > 0)$$

and

$$(1.24) \qquad \sigma(I,K^{(n)},f) = \int_I |K^{(n)}(\chi_I f)(x)| \ dx \qquad (n \geqq 1).$$

Then

$$\|K\|_{2,2} \leqq \text{Const} \ \{\sigma(K) + C_\delta\omega_\delta(K)\} \ \leqq \text{Const} \ \{\sigma(K) + C_\delta\} \ .$$

(See Lemma 2.5 in Chapter II.) For $n \geqq 1$, $f \in L^{\infty}_{real,1}$ and an interval I, we have

$$\frac{1}{|I|} \sigma(I,K^{(2^{n-1})},f) \leqq \{\frac{1}{|I|} \int_{-\infty}^{\infty}|K^{(2^{n-1})}(\chi_I f)(x)|^2 \ dx\}^{1/2}$$

$$= \{\frac{1}{|I|} \int_I \ f(x) \ K^{(2^n)}(\chi_I f)(x)dx\}^{1/2} \leqq \ \{ \frac{1}{|I|} \ \sigma(I,K^{(2^n)},f)\}^{1/2}$$

and hence

$$\frac{1}{|I|} \ \sigma(I,K,f) \leqq \{ \frac{1}{|I|} \ \sigma(I,K^{(2^n)},f)\}^{2^{-n}} .$$

For a while, we assume that $\chi_I f$ is a step function, saying $\chi_I f = \Sigma_{k=1}^N \alpha_k \chi_{I_k}$ ($|\alpha_k| \leqq 1$, I_k interval, $I_k \cap I_\ell = \emptyset$ $(k \neq \ell)$). Then Shwartz's inequality and Lemma 1.17 show that

$$\{ \frac{1}{|I|} \sigma(I,K^{(2^n)},f)\}^2 \leqq \frac{1}{|I|} |\int_I f(x)\, K^{(2^{n+1})} (\chi_I f)(x)dx|$$

$$\leqq \frac{1}{|I|} \sum_{k=1}^N |\int_I f(x)\, K^{(2^{n+1})} \chi_{I_k}(x)dx|$$

$$\leqq \frac{1}{|I|} \sum_{k=1}^N |\int_{I_k} K^{(2^{n+1})}\chi_{I_k} (x)\, dx| + \omega_\delta(K^{(2^{n+1})}) \frac{1}{|I|} \sum_{k=1}^N \int_{I-I_k} \{\int_{I_k} \frac{dy}{|x-y|} \}dx$$

$$\leqq \beta(K^{(2^{n+1})}) + \text{Const } N\ \omega_\delta(K^{(2^{n+1})}) \leqq \text{Const } N\ C_\delta^{2^{n+1}} ,$$

which shows that $(1/|I|)\, \sigma(I,K,f) \leqq \{\text{Const } N\, C_\delta^{2^{n+1}}\}^{2^{-n}}$. Letting n tend to infinity, we have $(1/|I|)\, \sigma(I,K,f) \leqq C_\delta$. Since this inequality holds for any $N \geqq 1$, α_k, I_k $(1 \leqq k \leqq N)$, we can remove the above assumption, i.e., $(1/|I|)\, \sigma(I,K,f) \leqq C_\delta$ for the given f. Taking the supremum over all $f \in L^\infty_{\text{real},1}$ and all intervals I, we have $\sigma(K) \leqq C_\delta$, which implies Theorem 1.12.

We now give the proof of Lemma 1.17. Assuming that $\omega_\delta(K) = 1$, we inductively show that

$$\beta(K^{(n)}) + \omega_{\delta-\varepsilon}(K^{(n)}) \leqq C_0^n \qquad (n \geqq 1),$$

where C_0 is a constant depending only on δ, ε, and is determined later. Since $K(x,y)$ is anti-symmetric and $\omega_\delta(K) = 1$, we have

$$\beta(K^{(1)}) + \omega_{\delta-\varepsilon}(K^{(1)}) \leqq 0 + \omega_\delta(K^{(1)}) = 1.$$

Suppose that the required inequality holds for $n-1$. For the sake of simplicity, we use, from now, C for various constants depending only on δ, ε. First we show that $\beta(K^{(n)}) \leqq CC_0^{n-1}$. For an interval I, we have

$$\int_I K^{(n)} \chi_I(x)dx = \int_{-\infty}^\infty \{\int_I \int_I K(x,s)K^{(n-1)}(s,y)\, dx\, dy\}\, ds$$

$$= \int_I \{\int_I \int_I\} + \int_{I^c} \{\int_I \int_I\} = L_1 + L_2 ,$$

$$|L_2| \leqq C_0^{n-1} \int_{I^c} (\int_I \frac{dx}{|x-s|})(\int_I \frac{dy}{|s-y|}) \,ds \leqq CC_0^{n-1} |I|$$

and

$$|L_1| \leqq \{ \int_I |K\chi_I(x)|^2dx\}^{1/2} \{\int_I |K^{(n-1)} \chi_I(x)|^2\, dx\}^{1/2} .$$

Since $K^{(n-1)}1 = 0$, we have

$$\int_I |K^{(n-1)}\chi_I(x)|^2\,dx = \int_I |K^{(n-1)}\chi_{I^c}(x)|^2\,dx$$

$$\leqq 2\int_I |K^{(n-1)}\chi_{I^{*c}}(x)|^2\,dx + 2\int_I |K^{(n-1)}\chi_{I^*-I}(x)|^2\,dx$$

$$\leqq 2\int_I |K^{(n-1)}\chi_{I^{*c}}(x)|^2 dx + 2\,C_0^{2n-2}\int_I \Big(\int_{I^*-I}\frac{dy}{|x-y|}\Big)^2 dx$$

$$\leqq 2\int_I |K^{(n-1)}\chi_{I^{*c}}(x)|^2\,dx + C\,C_0^{2n-2}|I| = 2\,L_{10} + C\,C_0^{2n-2}\,|I|,$$

where I^* is the double of I. Since

$$\Big|\int_I K^{(n-1)}\chi_{I^{*c}}(x)dx\Big| \leqq \Big|\int_I K^{(n-1)}\chi_{I^c}(x)\,dx\Big| + \Big|\int_I K^{(n-1)}\chi_{I^*-I}(x)dx\Big|$$

$$\leqq \Big|\int_I K^{(n-1)}\chi_I(x)dx\Big| + C_0^{n-1}\int_I\Big(\int_{I^*-I}\frac{dy}{|x-y|}\Big)\,dx$$

$$\leqq \beta(K^{(n-1)})\,|I| + C\,C_0^{n-1}\,|I| \leqq C\,C_0^{n-1}\,|I|,$$

we have

$$L_{10} \leqq 2\int_I |K^{(n-1)}\chi_{I^{*c}}(x) - (K^{(n-1)}\chi_{I^{*c}})_I|^2\,dx + C\,C_0^{2n-2}\,|I|$$

$$\leqq \frac{2}{|I|^2}\int_I \{\int_I (\int_{I^{*c}} |K^{(n-1)}(x,y)-K^{(n-1)}(s,y)|dy)ds\}^2\,dx + C\,C_0^{2n-2}\,|I|$$

$$\leqq (CC_0^{2n-2}/|I|^2)\int_I \{\int_I (\int_{I^{*c}} \frac{|x-s|^{\delta-\varepsilon}}{|x-y|^{1+\delta-\varepsilon}})\,ds\}^2\,dx + C\,C_0^{2n-2}\,|I|$$

$$\leqq CC_0^{2n-2}\,|I|.$$

Thus $\int_I |K^{(n-1)}\chi_I(x)|^2\,dx \leqq C\,C_0^{2n-2}\,|I|$. In the same manner, $\int_I |K\chi_I(x)|^2\,dx \leqq C|I|$, and hence $|L_1| \leqq CC_0^{n-1}\,|I|$. Consequently we have $|\int_I K\chi_I(x)dx| \leqq C\,C_0^{n-1}\,|I|$. Since I is arbitrary, we obtain $\beta(K^{(n)}) \leqq CC_0^{n-1}$.

Next we show that $\omega_{\delta-\varepsilon}(K^{(n)}) \leqq C\,C_0^{n-1}$. In the same manner as in the estimate of $|L_1|$, we have

$$\sup_{x\in\mathbb{R},\varepsilon>0}\Big|\int_{|x-y|>\varepsilon} K^{(n-1)}(x,y)dy\Big| \leqq C\,C_0^{n-1},$$

$$\sup_{x\in\mathbb{R},\varepsilon>0}\Big|\int_{|x-y|>\varepsilon} K(x,y)dy\Big| \leqq C.$$

For $x, y \in \mathbb{R}$, we have

$$K^{(n)}(x,y) = \int_{-\infty}^{\infty} K(x,s)K^{(n-1)}(s,y)ds$$

$$= \int_{I_1'} + \int_{I_2'} + \int_{I_3'} = L_1' + L_2' + L_3' ,$$

where I_1' is the interval of midpoint x and of length $|x-y|/2$, I_2' is the interval of midpoint y and of length $|x-y|/2$ and $I_3' = (I_1' \cup I_2')^c$. We have

$$|L_3'| \leq C_0^{n-1} \int_{I_3'} \frac{ds}{|x-s||s-y|} \leq C\, C_0^{n-1} /|x-y| ,$$

$$|L_2'| \leq |\int_{I_2'} \{K(x,s)-K(x,y)\}\, K^{(n-1)}(s,y)ds| + |K(x,y)||\int_{I_2'} K^{(n-1)}(s,y)ds|$$

$$\leq C\, C_0^{n-1} \int_{I_2'} \frac{|s-y|^{\delta}}{|x-y|^{1+\delta}} \frac{1}{|s-y|} ds + \frac{1}{|x-y|} C\, C_0^{n-1} \leq C\, C_0^{n-1}/|x-y|$$

and

$$|L_1'| \leq |\int_{I_1'} K(x,s) \{K^{(n-1)}(s,y)-K^{(n-1)}(x,y)\}ds| + |K^{(n-1)}(x,y)||\int_{I_1'}K(x,s)ds|$$

$$\leq C\, C_0^{n-1} \int_{I_1'} \frac{1}{|x-s|} \frac{|s-x|^{\delta-\varepsilon}}{|x-y|^{1+\delta-\varepsilon}} ds + \frac{C\, C_0^{n-1}}{|x-y|} C \leq C\, C_0^{n-1}/|x-y| .$$

Thus $|K^{(n)}(x,y)| \leq C\, C_0^{n-1}/|x-y|$. For $x, x', y \in \mathbb{R}$ with $|x-x'| \leq |x-y|/2$, we have

$$K^{(n)}(x,y) - K^{(n)}(x',y) = \int_{-\infty}^{\infty}\{K(x,s)-K(x',s)\}\, K^{(n-1)}(s,y)\, ds$$

$$= \sum_{k=1}^{6} \int_{I_k''} = \sum_{k=1}^{6} L_k'' ,$$

where $I_1'' = (x - \frac{|x-x'|}{2}, x + \frac{|x-x'|}{2})$, $I_2'' = (x' - \frac{|x-x'|}{2}, x' + \frac{|x-x'|}{2})$ $I_3'' = (x - \frac{3|x-y|}{4}, x + \frac{3|x-y|}{4}) - (I_1'' \cup I_2'')$, $I_4'' = (y - \frac{|x-x'|}{10}, y + \frac{|x-x'|}{10})$ $I_5'' = (y - \frac{|x-y|}{4}, y + \frac{|x-y|}{4}) - I_4''$ and $I_6'' = (I_1'' \cup\ldots\cup I_5'')^c$. We have

$$|L_6''| \leq C\, C_0^{n-1} \int_{I_6''} \frac{|x-x'|^{\delta}}{|x-s|^{1+\delta}} \frac{1}{|s-y|} ds \leq C\, C_0^{n-1}|x-x'|^{\delta}/|x-y|^{1+\delta}$$

$$\leq C\, C_0^{n-1}|x-x'|^{\delta-\varepsilon}/|x-y|^{1+\delta-\varepsilon} ,$$

$$|L_5''| \leq C\, C_0^{n-1} \frac{|x-x'|^\delta}{|x-y|^{1+\delta}} \int_{I_5''} \frac{ds}{|s-y|}$$

$$\leq C\, C_0^{n-1} \frac{|x-x'|^\delta}{|x-y|^{1+\delta}} \log\left(C \frac{|x-y|}{|x-x'|}\right) \leq C\, C_0^{n-1} |x-x'|^{\delta-\varepsilon}/|x-y|^{1+\delta-\varepsilon},$$

$$|L_4''| \leq \left|\int_{I_4''} \{K(x,s)-K(x',s)-K(x,y) + K(x',y)\}\, K^{(n-1)}(s,y)\, ds\right|$$

$$+ |K(x,y) - K(x',y)| \left|\int_{I_4''} K^{(n-1)}(s,y) ds\right|$$

$$\leq C\, C_0^{n-1} \int_{I_4''} \left\{ \frac{|s-y|^\delta}{|x-y|^{1+\delta}} + \frac{|s-y|^\delta}{|x'-y|^{1+\delta}} \right\} \frac{1}{|s-y|}\, ds$$

$$+ \frac{|x-x'|^\delta}{|x-y|^{1+\delta}}\, C\, C_0^{n-1} \leq C\, C_0^{n-1} |x-x'|^\delta/|x-y|^{1+\delta}$$

$$\leq C\, C_0^{n-1} |x-x'|^{\delta-\varepsilon}/|x-y|^{1+\delta-\varepsilon},$$

$$L_3'' = \int_{I_3''} \{K(x,s)-K(x',s)\} \{K^{(n-1)}(s,y) - K^{(n-1)}(x',y)\}\, ds$$

$$+ K^{(n-1)}(x',y) \int_{I_3''} \{K(x,s) - K(x',s)\}\, ds = L_{31}'' + L_{30}'',$$

$$L_2'' = \int_{I_2''} \{K(x,s)-K(x',s)\} \{K^{(n-1)}(s,y) - K^{(n-1)}(x',y)\}\, ds$$

$$+ K^{(n-1)}(x',y) \int_{I_2''} \{K(x,s)-K(x',s)\}\, ds = L_{21}'' + L_{20}''$$

and

$$L_1'' = \int_{I_1''} \{K(x,s)-K(x',s)\} \{K^{(n-1)}(s,y) - K^{(n-1)}(x,y)\}\, ds$$

$$+ \{K^{(n-1)}(x,y) - K^{(n-1)}(x',y)\} \int_{I_1''} \{K(x,s) - K(x',s)\}\, ds$$

$$+ K^{(n-1)}(x',y) \int_{I_1''} \{K(x,s) - K(x',s)\}\, ds = L_{11}'' + L_{12}'' + L''_{10}.$$

Since

$$|L_{31}''| \leq C\, C_0^{n-1} \int_{I_3''} \frac{|x-x'|^\delta}{|x'-s|^{1+\delta}} \frac{|s-x'|^{\delta-\varepsilon}}{|x-y|^{1+\delta-\varepsilon}}\, ds$$

$$\leq C\, C_0^{n-1} |x-x'|^{\delta-\varepsilon}/|x-y|^{1+\delta-\varepsilon},$$

$$|L_{21}''| \leq C\, C_0^{n-1} \int_{I_2''} \frac{1}{|x'-s|} \frac{|s-x'|^{\delta-\varepsilon}}{|x-y|^{1+\delta-\varepsilon}}\, ds$$

$$\leq C\, C_0^{n-1} |x-x'|^{\delta-\varepsilon}/|x-y|^{1+\delta-\varepsilon},$$

$$|L''_{11}| \leq C\, C_0^{n-1} \int_{I''_1} \frac{1}{|x-s|} \frac{|s-x|^{\delta-\varepsilon}}{|x-y|^{1+\delta-\varepsilon}}\, ds$$

$$\leq C\, C_0^{n-1} |x-x'|^{\delta-\varepsilon}/|x-y|^{1+\delta-\varepsilon}$$

and

$$|L''_{12}| \leq C_0^{n-1} \frac{|x-x'|^{\delta-\varepsilon}}{|x-y|^{1+\delta-\varepsilon}} \left\{ \left|\int_{I''_1} K(x,s)ds\right| + \int_{I''_1} \frac{ds}{|x'-s|} \right\}$$

$$\leq C\, C_0^{n-1} |x-x'|^{\delta-\varepsilon}/|x-y|^{1+\delta-\varepsilon} ,$$

we have, with $I''_0 = I''_1 \cup I''_2 \cup I''_3$,

$$|K^{(n)}(x,y) - K^{(n)}(x',y)| \leq |L''_{10} + L''_{20} + L''_{30}| + C\, C_0^{n-1} \frac{|x-x'|^{\delta-\varepsilon}}{|x-y|^{1+\delta-\varepsilon}}$$

$$= |K^{(n-1)}(x',y)| \left|\int_{I''_0} \{K(x,s)-K(x',s)\}ds\right| + C\, C_0^{n-1} \frac{|x-x'|^{\delta-\varepsilon}}{|x-y|^{1+\delta-\varepsilon}}$$

$$\leq C\, C_0^{n-1} \frac{1}{|x-y|} \left|\int_{I''^c_0} \{K(x,s)-K(x',s)\}ds\right| + C\, C_0^{n-1} \frac{|x-x'|^{\delta-\varepsilon}}{|x-y|^{1+\delta-\varepsilon}}$$

$$\leq C\, C_0^{n-1} \frac{1}{|x-y|} \int_{I''^c_0} \frac{|x-x'|^{\delta}}{|x-s|^{1+\delta}}\, ds + C\, C_0^{n-1} \frac{|x-x'|^{\delta-\varepsilon}}{|x-y|^{1+\delta-\varepsilon}}$$

$$\leq C\, C_0^{n-1} |x-x'|^{\delta-\varepsilon}/|x-y|^{1+\delta-\varepsilon} .$$

Since $K^{(n)}(x,y)$ is either anti-symmetric or symmetric, we have also

$$|K^{(n)}(x,y)- K^{(n)}(x,y')| \leq C\, C_0^{n-1} |y-y'|^{\delta-\varepsilon}/|x-y|^{1+\delta-\varepsilon}$$

for $x, y, y' \in \mathbb{R}$ with $|y-y'| \leq |x-y|/2$. Thus $\omega_{\delta-\varepsilon}(K^{(n)}) \leq C\, C_0^{n-1}$. Consequently, $\beta(K^{(n)}) + \omega_{\delta-\varepsilon}(K^{(n)}) \leq C\, C_0^{n-1}$. This shows that $\beta(K^{(n)}) + \omega_{\delta-\varepsilon}(K^{(n)}) \leq C_0^n$ if C_0 is large enough. This completes the proof of Lemma 1.17.

In Chapter I, we showed 8 proofs of the boundedness of the Calderón commutator $T[\cdot]$. Since the Calderón commutator is closely related to analycity of functions, it seems necessary to give more proofs and to have a unified understanding.

CHAPTER II. A REAL VARIABLE METHOD FOR THE CAUCHY TRANSFORM ON GRAPHS

§2.1. Coifman-McIntosh-Meyer's Theorem ([7])

For a real-valued locally integrable function a, we define a kernel by

$$(2.1)\qquad C[a](x,y) = 1/\{(x-y) + i(A(x) - A(y))\},$$

where A is a primitive of a. We write simply by $C[a]$ the singular integral operator defined by the kernel (2.1). This is called the Cauchy transform of Calderón on a graph $\{(x, A(x));\ x \in \mathbb{R}\}$. We put

$$L^\infty_{real} = \bigcup_{\beta > 0} L^\infty_{real,\beta} = \{a \in L^\infty;\ a \text{ is real-valued}\}.$$

(See (1.23).) Coifman-McIntosh-Meyer showed

Theorem B ([7]). The norm $\|C[a]\|_{2,2}$ is bounded if $a \in L^\infty_{real}$.

The operator $C[a]$ is expressed formally in the following form

$$C[a] = (-\pi)H + \sum_{n=1}^{\infty} (-i)^n T_n[a],$$

where $T_1[a] = T[a]$ (the Calderón commutator) and $T_n[a]$ is the n-th Coifman-Meyer commutator $(n \geqq 2)$, i.e., $T_n[a]$ is an operator defined by

$$(2.2)\qquad T_n[a](x,y) = (A(x) - A(y))^n/(x-y)^{n+1}.$$

Prior to this theorem, the following three theorems were shown. Calderón showed that $\|T_1[a]\|_{2,2} \leqq \text{Const}\ \|a\|_\infty$ $(a \in L^\infty)$, Coifman-Meyer [9] showed that

$$(2.3)\qquad \|T_n[a]\|_{2,2} \leqq \text{Const}\ n!\ \|a\|^n_\infty \quad (a \in L^\infty,\ n \geqq 2)$$

and Calderón showed that

$$(2.4)\qquad \|C[a]\|_{2,2} \text{ is bounded if } \|a\|_\infty\ (a \in L^\infty_{real}) \text{ is small enough.}$$

At present, there are three proofs of Theorem B; the original proof, a proof by the Tb theorem [40] and a proof by perturbation. In this chapter, we show a self-contained proof by perturbation. A proof by perturbation was first given by Calderón [4] and David [17]. Improving their methods and repeating a simple perturbation method, we shall deduce Theorem B only from the boundedness of H ([17], [42], [45]). (See APPENDIX II.)

§2.2. Two basic principles (Zygmund [54])

Here are two basic principles in real analysis.

Covering Lemma. Let $\{I_\lambda\}_{\lambda \in \Lambda}$ be a family of intervals in $\mathbb{R}$ such that $|\cup_{\lambda\in\Lambda} I_\lambda| < \infty$. Then there exists a sequence $\{I_{\lambda_k}\}_{k=1}^{\infty}$ of mutually disjoint intervals such that

$$|\cup_{\lambda\in\Lambda} I_\lambda| \leqq 5 \sum_{k=1}^{\infty} |I_{\lambda_k}|.$$

The proof is as follows. Let I_{λ_1} be an interval such that $2|I_{\lambda_1}|$ is larger than the supremum of $|I_\lambda|$ over all $\lambda \in \Lambda$. Suppose that $I_{\lambda_1}, \ldots, I_{\lambda_{k-1}}$ have been chosen. Let I_{λ_k} be an interval such that $2|I_{\lambda_k}|$ is larger than the supremum of $|I_\lambda|$ over all $\lambda \in \Lambda_{k-1}$, where $\Lambda_0 = \Lambda$ and $\Lambda_{k-1} = \{\lambda \in \Lambda\ ;\ I_\lambda \cap I_{\lambda_j} = \emptyset\ (1 \leqq j \leqq k-1)\}$ $(k \geqq 2)$. (If $\Lambda_{k-1} = \emptyset$, we stop our induction at $k-1$.)

Now we show that $\{I_{\lambda_k}\}$ is the required sequence. We first assume that $\{I_{\lambda_k}\}$ is an infinite sequence. Since the intervals are mutually disjoint and $|\cup_{k=1}^{\infty} I_{\lambda_k}| < \infty$, we have $\lim_{k\to\infty} |I_{\lambda_k}| = 0$. For I_λ, there exists I_{λ_j} such that $|I_\lambda| > 2|I_{\lambda_j}|$, which implies that $\lambda \notin \Lambda_j$. Hence $\{j;\ \lambda \notin \Lambda_j\} \neq \emptyset$. Let k be the smallest integer in the set. Then $|I_\lambda| \leqq 2|I_{\lambda_k}|$, according to the definition of our choice. Since $\lambda \notin \Lambda_k$, we have $I_\lambda \cap I_{\lambda_k} \neq \emptyset$, which gives that $I_\lambda \subset \tilde{I}_{\lambda_k}$, where $\tilde{I}_{\lambda_k}$ is the interval of the same midpoint as I_{λ_k} and of length $5|I_{\lambda_k}|$. Thus

$$|\cup_{\lambda\in\Lambda} I_\lambda| \leqq |\bigcup_{k=1}^{\infty} \tilde{I}_{\lambda_k}| \leqq 5 \sum_{k=1}^{\infty} |I_{\lambda_k}|.$$

If $\{I_{\lambda_k}\}$ is a finite sequence, each I_λ intersects with $\bigcup I_{\lambda_k}$. Hence, in the same manner, we have the required inequality. This completes the proof of this lemma.

Rising Sun Lemma. Let a be a function in an interval I such that $\alpha \leqq a(x) \leqq \beta$ for any $x \in I$, where $\beta \geqq 0$. Let A be a primitive of a. For γ $(\alpha \leqq \gamma \leqq \beta)$, we define a function B in I by $B(x) = \inf \Phi(x)$, where the infimum is taken over all functions Φ such that $\Phi \geqq A$, $\Phi' \geqq \gamma$ a.e. on I.

Let $b = B'$ and $\Omega = \{x \in I;\ A(x) \neq B(x)\} = \cup_{k=1}^{\infty} I_k$, where $\{I_k\}_{k=1}^{\infty}$ are the components of Ω. Then

(2.5) $\gamma \leq b(x) \leq \beta$ a.e. on I,

(2.6) $b(x) = \gamma \quad (x \in \Omega)$,

(2.7) $(a)_{I_k} \leq \gamma \quad ((a)_{I_k} = \frac{1}{|I_k|} \int_{I_k} a(s)ds,\ k \geq 1)$,

(2.8) $|\Omega| \leq \frac{\beta - (b)_I}{\beta - \gamma} |I| \quad ((b)_I = \frac{1}{|I|} \int_I b(s)ds)$.

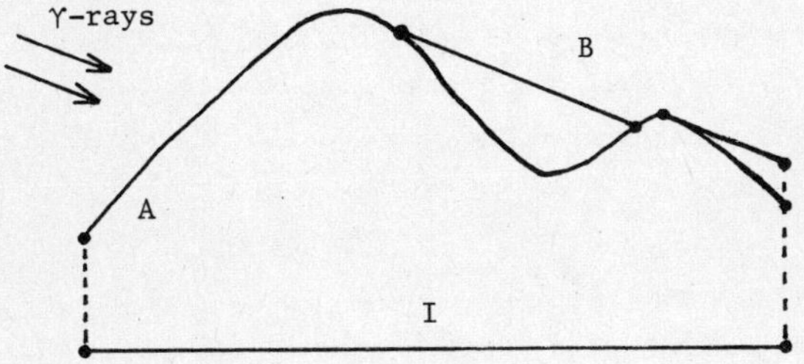

Inequalities (2.5)-(2.7) are easily seen. We have

$$(b)_I |I| = \int_I b(s)ds = \int_{I-\Omega} + \int_{\Omega}$$

$$\leq \beta|I - \Omega| + \gamma|\Omega| = \beta|I| - (\beta - \gamma) |\Omega| ,$$

which gives (2.8). For the sake of convenience, we call this rising sun lemma RSL of Type 1 (γ-*ray*, β-*ascent*); we shall use later various rising sun lemmas. For an open set Ω, we denote by $\{I_{\Omega,k}\}_{k=1}^{\infty}$ its components. The following two lemmas are also the rising sun lemmas for integrable functions.

Lemma 2.1 (The Calderón-Zygmund decomposition [35, p. 12]). Let $f \in L^1$ and $\lambda > 0$. Then there exists an open set Ω such that

$$|\Omega| \leq \|f\|_1/\lambda,\ (|f|)_{I_{\Omega,k}} = \lambda \quad (k \geq 1),\quad |f(x)| \leq \lambda \quad \text{a.e. on} \quad \Omega^c.$$

To see this, we put $A(x) = \int_0^x |f(s)|ds \quad (x > 0)$, and define a function B in $(0,\infty)$ by $B(x) = \sup \Phi(x)$, where the supremum is taken over all functions Φ such that $\Phi \leq A$, $\Phi' \leq \lambda$ a.e. on $(0,\infty)$. Let $\Omega_1 = \{x > 0;\ A(x) \neq B(x)\}$. Then

$$|\Omega_1| \leq \frac{1}{\lambda} \int_0^{\infty} |f(s)|ds, \qquad (|f|)_{I_{\Omega_1,k}} = \lambda \quad (k \geq 1),$$

$$|f(x)| \leq \lambda \quad \text{a.e. on} \quad (0,\infty) - \Omega_1 .$$

Considering $f(-x)$, we obtain, in the same manner, an open set Ω_2 in $(0,\infty)$. Then $\Omega_1 \cup \{-x;\ x \in \Omega_2\}$ is the required open set.

In the same manner, we have

Lemma 2.2. Let f be an integrable function in an interval I and let $\lambda > 0$ satisfy $\lambda > (|f|)_I$. Then there exists an open set Ω in I such that

$$|\Omega| \leq \frac{1}{\lambda} \int_I |f(x)|ds, \quad (|f|)_{I_{\Omega,k}} \leq \lambda \quad (k \geq 1),$$

$$|f(x)| \leq \lambda \quad \text{a.e. on } I - \Omega.$$

The (non-centered) maximal operator M is defined by $Mf(x) = \sup(|f|)_I$, where the supremum is taken over all intervals I containing x. For $p > 1$, $\|M\|_{p,p}$ denotes the norm of M as an operator from L^p to itself. The following lemma is deduced from Covering Lemma.

Lemma 2.3 ([35, p.7]). $\|M\|_{p,p} \leq C_p$ $(p > 1)$.

For $f \in L^1$, $\lambda > 0$, we put $E_\lambda = \{x;\ Mf(x) > \lambda\}$. For each $x \in E_\lambda$, we can choose an interval I_x containing x so that $(|f|)_{I_x} > \lambda$. Covering Lemma shows that there exists a sequence $\{I_{x_k}\}_{k=1}^{\infty}$ of mutually disjoint intervals such that $|E_\lambda| \leq 5\ \Sigma_{k=1}^{\infty} |I_{x_k}|$, which yields that

$$|x;\ Mf(x) > \lambda| \leq \frac{5}{\lambda} \sum_{k=1}^{\infty} \int_{I_k} |f(s)|ds \leq \frac{5}{\lambda} \|f\|_1 .$$

For $f \in L^p$ and $\lambda > 0$, we define f_λ by $f_\lambda(x) = f(x)$ if $|f(x)| > \lambda/2$ and $f_\lambda(x) = 0$ if $|f(x)| \leq \lambda/2$. Then

$$\|M f\|_p^p = C_p \int_0^\infty \lambda^{p-1}|x;\ Mf(x) > \lambda|dx$$

$$\leq C_p \int_0^\infty \lambda^{p-1} \{|x;\ Mf_\lambda(x) > \lambda/2| + |x;\ M(f-f_\lambda)(x) > \lambda/2\ |\}\ d\lambda$$

$$= C_p \int_0^\infty \lambda^{p-1}\ |x;\ Mf_\lambda(x) > \lambda/2|\ d\lambda$$

$$\leq C_p \int_0^\infty \lambda^{p-2}\ \|f_\lambda\|_1\ d\lambda = C_p \int_0^\infty \lambda^{p-2}\ \{\int_{\lambda/2}^\infty |x;|f(x)| > s\ |\ ds\}\ d\lambda$$

$$= C_p \int_0^\infty |x;|f(x)| > s|\ \{\int_0^{2s} \lambda^{p-2}\ d\lambda\}\ ds = C_p\ \|f\|_p^p ,$$

which gives that $\|M\|_{p,p} \leq C_p$.

At last we note John-Nirenberg's inequality, which was used in Chapter I. This is deduced from RSL. (For the proof of Theorem B, this is not necessary.)

Lemma 2.4 ([32]). Let $f \in BMO$ and I be an interval. Then

$$|x \in I;\ |f(x)-(f)_I| > \lambda| \leqq \exp(-\text{ Const }\lambda)|I| \qquad (\lambda \geqq 1).$$

§2.3. σ-function ([8], [35], [54])

In this section, we show a fundamental inequality for standard kernels. For the sake of simplicity, we deal with only kernels $K(x,y)$ satisfying (1.21). (See §1.9.) We use the notation $\sigma(K)$, $\sigma(I,K,f)$ in (1.22), (1.24). For a standard kernel, we define an operator K^* by

$$K^*f(x) = \sup_{\varepsilon > 0} \left|\int_{|x-y| > \varepsilon} K(x,y)f(y)dy\right| .$$

We show

Lemma 2.5 ([35], p. 49). Let $K(x,y)$ be a δ-standard kernel (satisfying (1.21)). Then $\|K^*\|_{2,2} \leqq \text{Const}\ \sigma(K) + C_\delta\ \omega_\delta(K)$.

We begin by showing

$$(2.9) \qquad \sigma(K^*) \leqq \text{Const}\ \sigma(K) + C_\delta\ \omega_\delta(K),$$

where $\sigma(K^*)$ is the supremum of $(1/|I|) \int_I K^*(\chi_I f)(x)dx$ over all $f \in L^\infty_{real,1}$ and intervals I. For $\varepsilon > 0$, $f \in L^\infty_{real,1}$, an interval I and a point x on I, we put $J' = (x - \varepsilon/2, x + \varepsilon/2)$, $J = (x - \varepsilon, x + \varepsilon)$, $g = \chi_{I \cap J} f$ and $h = \chi_{I-J} f$. If $0 < \varepsilon < |I|$, we have, for any $s \in J'$,

$$\begin{aligned}\left|\int_{|x-y| > \varepsilon} K(x,y)(\chi_I f)(y)dy\right| &= |Kh(x)| \leqq |Kh(s)| + |Kh(x)-Kh(s)| \\ &\leqq |Kh(s)| + C_\delta\ \omega_\delta(K) \leqq |K(\chi_I f)(s)| + |Kg(s)| + C_\delta\ \omega_\delta(K) \\ &= |\chi_{I^*}(s)K(\chi_I f)(s)| + |Kg(s)| + C_\delta\ \omega_\delta(K),\end{aligned}$$

where I^* is the double of I. Taking first the square roots of the first quantity and the last three quantities, and taking next their means over J' with respect to s, we obtain

$$\begin{aligned}&\left|\int_{|x-y| > \varepsilon} K(x,y)(\chi_I f)(y)dy\right|^{1/2} \\ &\qquad \leqq M(|\chi_{I^*} K(\chi_I f)|^{1/2})(x) + (|Kg|^{1/2})_{J'} + C_\delta\ \omega_\delta(K)^{1/2}.\end{aligned}$$

If $\varepsilon \geqq |I|$, then $\int_{|x-y| > \varepsilon} K(x,y)(\chi_I f)(y)dy = 0$. Hence this inequality holds for all $\varepsilon > 0$, which shows that this inequality holds with the first quantity replaced by $K^*(\chi_I f)(x)^{1/2}$. Taking the squares of both sides of the resulting inequality, and using Shwartz's inequality, we obtain

$$K^*(\chi_I f)(x) \leq \text{Const}\ \ M(|\chi_{I^*}\ K(\chi_I f)|^{1/2})(x)^2$$
$$+ \text{Const}(|Kg|^{1/2})^2_{J'} + C_\delta\ \omega_\delta(K).$$

Since

$$(\ M(|\chi_{I^*} K(\chi_I f)|^{1/2})\ ^2)_I \leq \text{Const}\ \frac{1}{|I|}\ \int_{I^*} |K(\chi_I f)(x)|dx$$

$$\leq\ \text{Const}\ \frac{1}{|I|}\ \{\sigma(I,K,f) + \omega_\delta(K) \int_{I^*-I} (\int_I \frac{dy}{|x-y|})\ dx\}$$

$$\leq\ \text{Const}\ \ \{\sigma(K) + \omega_\delta(K)\}$$

and

$$(|Kg|^{1/2})^2_{J'} \leq (|Kg|)_{J'} \leq (|K(\chi_{J'} g)|)_{J'} + (|K(\chi_{J-J'} g)|)_{J'}$$

$$\leq\ \sigma(K) + \text{Const}\ \ \omega_\delta(K),$$

we have $(K^*(\chi_I f))_I \leq \text{Const}\ \ \sigma(K) + C_\delta\ \omega_\delta(K)$, which implies (2.9).

Let $f \in L^2$, $\lambda > 0$. We show the following good λ inequality:

$$(2.10)\qquad |x;\ K^*f(x) > 3\lambda,\ \ Mf(x) \leq \eta\lambda| \ \leq\ \frac{1}{10}\ |x;\ K^*f(x) > \lambda|\ ,$$

where $\eta > 0$ is determined later. To prove this, it is sufficient to show that, for each component I of $\{x;\ K^*f(x) > \lambda\}$,

$$|x \in I;\ K^*f(x) > 3\lambda,\ \ Mf(x) \leq \eta\lambda| \leq \frac{1}{10}\ |I|.$$

If $Mf(x) > \eta\lambda$ on I, this inequality evidently holds. Assuming that $M\,f(\xi) \leq \eta\lambda$ for some $\xi \in I$, we prove

$$(2.11)\qquad |x \in I;\ K^*f(x) > 3\lambda| \leq \frac{1}{10}\ |I|.$$

(See §1.4). Let $g = \chi_J f$ and $h = \chi_{J^c} f$, where $J = (x_0 - 2|I|,\ x_0 + 2|I|)$ (x_0 is the left endpoint of I). Then we have

$$|x \in I;\ K^*f(x) > 3\lambda| \leq\ |x \in I;\ K^*g(x) > \lambda|$$
$$+\ |x \in I;\ K^*h(x) > 2\lambda|\ \ \ (= L_1 + L_2,\ \text{say}).$$

First we estimate L_2. Note that $K^*h(x_0) \leq \lambda$. For $\varepsilon > 0$ and $x \in I$, we have

$$|\int_{|x-y|>\varepsilon} K(x,y)h(y)dy - \int_{|x_0-y|>\varepsilon} K(x_0,y)h(y)dy|$$

$$\leq \int_{-\infty}^{\infty} |K(x,y) - K(x_0,y)||h(y)|\, dy + \text{Const}\ \omega_\delta(K)\, Mf(\xi)$$

$$\leq C_\delta\, \omega_\delta(K)\, Mf(\xi) \leq C_\delta\, \omega_\delta(K)\, \eta\lambda.$$

Since $\varepsilon > 0$ is arbitrary, we have, with a constant $C_{\delta,1}$ depending only on δ,

$$\text{(2.12)} \qquad K^*h(x) \leq K^*h(x_0) + C_{\delta,1}\, \omega_\delta(K)\, \eta\lambda$$

$$\leq \{1 + C_{\delta,1}\, \omega_\delta(K)\, \eta\}\lambda \qquad (x \in I).$$

This shows that $L_2 = 0$ if $C_{\delta,1}\, \omega_\delta(K)\, \eta < 1$. Next we estimate L_1. By Lemma 2.1, there exists an open set $\Omega = \cup_{k=1}^{\infty} I_k$ $(I_k = I_{\Omega,k})$ such that

$$|\Omega| \leq \|g\|_1/(100\,\eta\lambda), \ (|g|)_{I_k} = 100\,\eta\lambda \qquad (k \geq 1),$$

$$|g(x)| \leq 100\,\eta\lambda \quad \text{a.e. on } \Omega^c.$$

We define a function $\tilde{g}$ by

$$\tilde{g}(x) = \begin{cases} g(x) & (x \in \Omega^c), \\ (g)_{I_k} & (x \in I_k,\ k \geq 1). \end{cases}$$

Then $\|\tilde{g}\|_\infty \leq 100\eta\lambda$. Put $\Omega^* = \cup_{k=1}^{\infty} I_k^*$, where I_k^* is the double of I_k. Then

$$|\Omega^*| \leq 2|\Omega| \leq \|g\|_1/(50\,\eta\lambda) \leq Mf(\xi)|J|/(50\,\eta\lambda) \leq |I|/15.$$

For $\varepsilon > 0$ and $x \in \Omega^{*c}$, there exist at most two intervals (saying I_1 and I_2) which intersect with the boundary of $(x - \varepsilon, x + \varepsilon)$. We have, with x_k = (the midpoint of I_k),

$$|\int_{|x-y|>\varepsilon} K(x,y)(g(y) - \tilde{g}(y))dy| = |\int_{(I_1 \cap I_2) \cap (x-\varepsilon,x+\varepsilon)^c} K(x,y)(g(y)-\tilde{g}(y))dy$$

$$+ \sum_{I_k \subset (x-\varepsilon,\ x+\varepsilon)^c} \int_{I_k} \{K(x,y)-K(x,x_k)\}\,(g(y) - \tilde{g}(y))dy|$$

$$\leq \text{Const}\ \omega_\delta(K)\ \{(|g|)_{I_1} + (|g|)_{I_2}\}$$

$$+ C_\delta\ \omega_\delta(K) \sum_{k=1}^{\infty} \frac{|I_k|^{1+\delta}}{(|x-x_k| + |I_k|)^{1+\delta}} (|g|)_{I_k} \leq C_\delta\ \omega_\delta(K)\ \eta\lambda\ (1 + \Delta(x)).$$

where $\Delta(x) = \Sigma_{k=1}^{\infty} |I_k|^{1+\delta}/(|x-x_k| + |I_k|)^{1+\delta}$. Since $\varepsilon > 0$ is arbitrary, we have

$$K^* g(x) \leq K^* \tilde{g}(x) + C_\delta \omega_\delta(K)\eta\lambda(1 + \Delta(x)) \qquad (x \in \Omega^{*c}).$$

Since $\operatorname{supp}(g) \subset J$ and $|\Omega| \leq |I|$, the support of $\tilde{g}$ is contained in the double J^* of J. Hence (2.9) shows that

$$\int_{I-\Omega^*} K^* \tilde{g}(x)dx \leq \int_{J^*} K^*(\chi_{J^*} \tilde{g})(x)dx$$

$$\leq \sigma(K^*)\ \|\tilde{g}\|_\infty\ |J^*| \leq \{\text{Const}\ \sigma(K) + C_\delta\ \omega_\delta(K)\}\ \eta\lambda\ |I|.$$

We have easily

$$\int_{I-\Omega^*} \{C_\delta\ \omega_\delta(K)\ \eta\lambda\ (1 + \Delta(x))\}\ dx$$

$$\leq C_\delta\ \omega_\delta(K)\ \eta\lambda\ \{|I| + |\Omega|\} \leq C_\delta\ \omega_\delta(K)\ \eta\lambda\ |I|.$$

Consequently, we have, with an absolute constant C_0 and a constant $C_{\delta,2}$ depending only on δ,

$$(2.13) \qquad L_1 \leq |x \in I - \Omega^*;\ K^* g(x) > \lambda| + |\Omega^*|$$

$$\leq \frac{1}{\lambda} \int_{I-\Omega^*} K^* g(x)dx + |I|/15$$

$$\leq \frac{1}{\lambda} \int_{I-\Omega^*} \{K^* \tilde{g}(x) + C_\delta\ \omega_\delta(K)\eta\lambda\ (1 + \Delta(x))\}\ dx + |I|/15$$

$$\leq \{(C_0\ \sigma(K) + C_{\delta,2}\ \omega_\delta(K))\eta + (1/15)\}\ |I|.$$

Let

$$\eta = \min\ \{(2\ C_{\delta,1}\omega_\delta(K))^{-1},\ (30\ C_0\ \sigma(K) + 30\ C_{\delta,2}\ \omega_\delta(K))^{-1}\}.$$

Then (2.12) and (2.13) show that

$$|x \in I;\ K^* f(x) > 3\lambda| \leq L_1 + L_2 = L_1 \leq \frac{1}{10}\ |I|.$$

Thus (2.11) holds, which implies (2.10).

In the same manner as in §1.4, (2.10) yields that

$$\|K^* f\|_2 \leq (\text{Const}/\eta)\ \|Mf\|_2 \leq \{\text{Const}\ \sigma(K) + C_\delta\ \omega_\delta(K)\}\ \|f\|_2,$$

which implies the required inequality in our lemma.

§2.4. A-priori estimates

In this section, we show some inequalities which play important roles later. For an operator T from L^2 to itself, we put

$$(2.14)\quad \tilde{\sigma}_0(T) = \sup \{\frac{1}{|I|}\ \tilde{\sigma}(I, T, \chi_I);\ I \ \text{interval}\},$$

$$(2.15)\quad \tilde{\sigma}(T) = \sup \{\frac{1}{|I|}\ \tilde{\sigma}(I, T, f);\ f \in L^{\infty}_{real,1},\ I \ \text{interval}\},$$

$$(2.16)\quad \hat{\sigma}(T) = \sup \{\frac{1}{|I|}\ \hat{\sigma}(I, T, f);\ 0 \leq f \leq 1,\ I \ \text{interval}\},$$

where

$$(2.17)\quad \tilde{\sigma}(I, T, f) = \int_I |T(\chi_I f)(x)|^2\, dx,$$

$$(2.18)\quad \hat{\sigma}(I, T, f) = \int_I |T(\chi_I f)(x)|^2\, f(x)dx.$$

For an open set Ω with $|\Omega| < \infty$, we put

$$(2.19)\quad \tilde{\sigma}(T;\Omega) = \sup \{\frac{1}{|I|}\ \tilde{\sigma}(I,T,f);\ f \in L^{\infty}_{real,1},\ I \ \text{component of} \ \Omega\}.$$

For a δ-standard kernel $K(x,y)$, we have easily $\sigma(K) \leq \tilde{\sigma}(K)^{1/2} \leq \|K\|_{2,2}$, and hence, by Lemma 2.5,

$$(2.20)\quad \tilde{\sigma}(K)^{1/2} \leq \|K\|_{2,2} \leq \text{Const}\ \tilde{\sigma}(K)^{1/2} + C_\delta\ \omega_\delta(K).$$

For a non-negative measure μ on $\mathbb{R}$, we denote by $(\cdot,\cdot)_\mu$ the inner product with respect to μ, i.e., $(f,g)_\mu = \int_{-\infty}^{\infty} f\bar{g}\, d\mu$. (In the case of the 1-dimension Lebesgue measure, we omit the suffix.) Here is an inequality necessary for the proof of Theorem B.

Lemma 2.6. Let I be an (open) interval, $\Omega = \cup_{k=1}^{\infty} I_k$ $(I_k = I_{\Omega,k})$ be an open set in I and let $K(x,y)$, $T(x,y)$ be two δ-standard kernels such that $K(x,y) = T(x,y)$ for any $x, y \in I - \Omega$, $x \neq y$. Then, for any $u, v \in L^2$ supported on $\bar{I}$ and a non-negative measure μ with $d\mu/dx \in L^{\infty}_{real,1}$,

$$|(Ku,v)_\mu| \leq |(Tu,v)_\mu| + \sum_{k=1}^{\infty} |((K-T)(\chi_{I_k} u), \chi_{I_k} v)_\mu|$$

$$+ C_\delta(\omega_\delta(K) + \omega_\delta(T))\ \|u\|_{*\Omega 2}\ \|v\|_{*\Omega 2},$$

where

$$(2.21)\quad \|w\|_{*\Omega 2} = \|w\|_2 + \{\sum_{k=1}^{\infty} \|\chi_{I_k^*} w\|_2^2\}^{1/2} \quad (w = u,v;\ I_k^* \text{ is the double of } I_k).$$

Proof. Since $\chi_{I-\Omega}(K-T)(\chi_{I-\Omega} u) = 0$, we have

$$|(Ku,v)_\mu| \leq |(Tu,v)_\mu| + |((K-T)(\chi_\Omega u), \chi_\Omega v)_\mu|$$

$$+ |((K-T)(\chi_\Omega u), \chi_{I-\Omega} v)_\mu| + |((K-T)(\chi_{I-\Omega} u), \chi_\Omega v)_\mu|$$

$$(= |(Tu,v)_\mu| + L_1 + L_2 + L_3, \text{ say}).$$

Without loss of generality, we may assume that $|I_1| \geq |I_2| \geq \dots$. We have, with $\chi_k = \chi_{I_k}$ $(k \geq 1)$,

$$L_1 \leq \sum_{k=1}^{\infty} |((K-T)(\chi_k u), \chi_k v)_\mu|$$

$$+ \sum_{k=1}^{\infty} \sum_{j=k+1}^{\infty} |((K-T)(\chi_k u), \chi_j v)_\mu| + \sum_{k=2}^{\infty} \sum_{j=1}^{k-1} |((K-T)(\chi_k u), \chi_j v)_\mu|$$

$$(= \sum_{k=1}^{\infty} |((K-T)(\chi_k u), \chi_k v)_\mu| + L_{11} + L_{12}, \text{ say}).$$

Let x_k be an endpoint of I_k such that $x_k \in I - \Omega$ $(k \geq 1)$. If $x \in I_k$, $y \in (I_k^{*c}) \cap I_j$, $k < j$, then

$$|K(x,y) - T(x,y)| = |K(x,y) - K(x_k,x_j) + T(x_k,x_j) - T(x,y)|$$

$$\leq C_\delta \, |I_k|^\delta / |x-y|^{1+\delta}.$$

(Here we assume $x_j \in (I_k^*)^c$. If $x_j \in I_k^*$, then we have evidently $|K(x,y) - T(x,y)| \leq C_\delta \, |I_k|^\delta / |x-y|^{1+\delta}$.) Hence

$$L_{11} \leq \sum_{k=1}^{\infty} \int_{I_k^*-I_k} |(K-T)(\chi_k u)(x) v(x)|dx$$

$$+ \sum_{k=2}^{\infty} \sum_{j=k+1}^{\infty} \int_{(I_k^{*c}) \cap I_j} |(K-T)(\chi_k u)(x) v(x)|dx$$

$$\leq C_\delta(\omega_\delta(K) + \omega_\delta(T)) \quad [\sum_{k=1}^{\infty} \int_{I_k^*-I_k} \{\int_{I_k} \frac{|u(y)|}{|x-y|} dy\} \; |v(x)| \; dx$$

$$+ \sum_{k=2}^{\infty} \int_{I_k^{*c}} \{\int_{I_k} \frac{|I_k|^\delta}{|x-y|^{1+\delta}} |u(y)|dy\} \; |v(x)| \; dx \,]$$

$$\leq C_\delta(\omega_\delta(K) + \omega_\delta(T)) \; \{ \sum_{k=1}^{\infty} \int_{I_k^*-I_k} |H(\chi_k u)(x) v(x)| \; dx$$

$$+ \sum_{k=2}^{\infty} \int_{I_k} |u(y)| \; M v(y) dy\} \leq C_\delta(\omega_\delta(K) + \omega_\delta(T)) \; \|u\|_2 \; \|v\|_{*\Omega 2}$$

$$\leq C_\delta(\omega_\delta(K) + \omega_\delta(T)) \; \|u\|_{*\Omega 2} \; \|v\|_{*\Omega 2} \, .$$

Using the adjoint kernel of $K(x,y) - T(x,y)$, we have, in the same manner,

$$L_{12} \leqq C_\delta(\omega_\delta(K) + \omega_\delta(T)) \|u\|_{*\Omega 2} \|v\|_{*\Omega 2} .$$

If $x \in I_k$, $y \in (I_k^{*c}) \cap (I - \Omega)$, then

$$|K(x,y) - T(x,y)| = |K(x,y) - K(x_k,y) + T(x_k,y) - T(x,y)|$$

$$\leqq C_\delta \; |I_k|^\delta / |x-y|^{1+\delta} .$$

Hence we have, in the same manner as in L_{11},

$$L_2 \leqq \sum_{k=1}^{\infty} \int_{I_k^*-I_k} |(K-T)(\chi_k u)(x)v(x)|dx$$

$$+ \sum_{k=1}^{\infty} \int_{(I_k^*)^c \cap (I-\Omega)} |(K-T)(\chi_k u)(x)v(x)| \, dx$$

$$\leqq C_\delta(\omega_\delta(K) + \omega_\delta(T)) \|u\|_{*\Omega 2} \|v\|_{*\Omega 2} .$$

Using the adjoint kernel, we have also

$$L_3 \leqq C_\delta(\omega_\delta(K)+ \omega_\delta(T)) \|u\|_{*\Omega 2} \|v\|_{*\Omega 2} .$$

Thus the required inequality holds. Q.E.D.

The following three lemmas are corollaries of Lemma 2.6.

Lemma 2.7. Let I, Ω, $K(x,y)$ and $T(x,y)$ be the same as in Lemma 2.6. Then, for any $f \in L^\infty_{real,1}$,

$$\sigma(I,K,f) \leqq \sigma(I,T,f) + \sum_{k=1}^{\infty} \sigma(I_k,K,f)$$

$$+ C_\delta(\tilde{\sigma}(T;\Omega)^{1/2} + \omega_\delta(K) + \omega_\delta(T)) \; |I|.$$

Proof. Using Lemma 2.6 with $u = \chi_I f$, $v = \chi_I \, K(\chi_I f)/|K(\chi_I f)|$ and $d\mu = dx$, we have

$$\sigma(I,K,f) = |(Ku,v)| \leqq |(Tu,v)| + \sum_{k=1}^{\infty} |((K-T)(\chi_k u), \chi_k v)|$$

$$+ C_\delta(\omega_\delta(K) + \omega_\delta(T)) \; |I| \leqq \sigma(I,T,f) + \sum_{k=1}^{\infty} \sigma(I,K,f)$$

$$+ C_\delta(\tilde{\sigma}(T;\Omega)^{1/2} + \omega_\delta(K) + \omega_\delta(T)) \; |I|.$$

Q.E.D.

Lemma 2.8. Let I, Ω be the same as in Lemma 2.6. Let $K(x,y)$, $T(x,y)$ be two anti-symmetric δ-standard kernels such that $K(x,y) = T(x,y)$

$(x, y \in I - \Omega,\ x \neq y)$. Then, for any $f \in L^{\infty}_{real,1}$ with $0 \leq f \leq 1$,

$$\hat{\sigma}(I,K,f) \leq \hat{\sigma}(I,T,f) + \sum_{k=1}^{\infty} \hat{\sigma}(I_k,K,f) + C_\delta\, \Lambda_1(K,T;\Omega)\,|I|$$

where

$$\Lambda_1(K,T;\Omega) = \{\tilde{\sigma}(T;\Omega)^{1/2} + \omega_\delta(K) + \omega_\delta(T)\}\,\{\sigma(K) + \sigma(T) + \omega_\delta(K) + \omega_\delta(T)\}\ .$$

Proof. Without loss of generality we may assume that $\operatorname{supp}(f) \subset I$. Using Lemma 2.6 with $u = f$, $v = \chi_I Kf$ and $d\mu = f\,dx$, we have

$$\hat{\sigma}(I,K,f) \leq |(Tf, \chi_I Kf)_{fdx}| + \sum_{k=1}^{\infty} |((K-T)(\chi_k f), \chi_k Kf)_{fdx}|$$

$$+ C_\delta(\omega_\delta(K) + \omega_\delta(T))\, \|f\|_{*\Omega 2}\, \|\chi_I Kf\|_{*\Omega 2} \quad (= L_1 + L_2 + L_3,\ \text{say}).$$

We have easily

$$L_3 \leq C_\delta(\omega_\delta(K) + \omega_\delta(T))\, \|K\|_{2,2}\, |I|$$

$$\leq C_\delta(\omega_\delta(K) + \omega_\delta(T))(\sigma(K) + \omega_\delta(K))|I| \leq C_\delta\, \Lambda_1(K,T;\Omega)\, |I|.$$

(See (2.22)). Since $K(x,y)$, $T(x,y)$ are anti-symmetric, $(\chi_k f(K-T)(\chi_k f))_{I_k} = 0$ $(k \geq 1)$. Hence we have, with $x_k' =$ (the midpoint of I_k),

$$L_2 = \sum_{k=1}^{\infty} \left|\int_{I_k} f(x)(K-T)(\chi_k f)(x)\, \overline{K\{(\chi_k + \chi_{I_k^*-I_k} + \chi_{I_k^{*c}})f\}(x)}\,dx\right|$$

$$\leq \sum_{k=1}^{\infty} \left|\int_{I_k} f(x)(K-T)(\chi_k f)(x)\, \overline{K(\chi_k f)(x)}\, dx\right|$$

$$+ \omega_\delta(K) \sum_{k=1}^{\infty} \int_{I_k} |(K-T)(\chi_k f)(x)| \left(\int_{I_k^*-I_k} \frac{dy}{|x-y|}\right) dx$$

$$+ \sum_{k=1}^{\infty} \left|\int_{I_k} f(x)(K-T)(\chi_k f)(x)\, \overline{\{ K(\chi_{I_k^{*c}} f)(x) - K(\chi_{I_k^{*c}} f)(x_k')\}}\, dx\right|$$

$$\leq \sum_{k=1}^{\infty} \hat{\sigma}(I_k,K,f) + \sum_{k=1}^{\infty} \|\chi_k T(\chi_k f)\|_2\, \|K(\chi_k f)\|_2$$

$$+ \omega_\delta(K) \sum_{k=1}^{\infty} \|(K-T)(\chi_k f)\|_2 \left\{\int_{I_k} \left(\int_{I_k^*-I_k} \frac{dy}{|x-y|}\right)^2 dx\right\}^{1/2}$$

$$+ C_\delta\, \omega_\delta(K) \sum_{k=1}^{\infty} \int_{I_k} |(K-T)(\chi_k f)(x)|\, dx$$

$$\leq \sum_{k=1}^{\infty} \hat{\sigma}(I_k,K,f) + C_\delta(\tilde{\sigma}(T;\Omega)^{1/2} + \omega_\delta(K))\,(\|K\|_{2,2} + \|T\|_{2,2})\, |I|$$

$$\leq \sum_{k=1}^{\infty} \hat{\sigma}(I_k,K,f) + C_\delta\, \Lambda_1(K,T;\Omega)\, |I|.$$

Using Lemma 2.6 with $u = f$, $v = \chi_I Tf$ and $d\mu = f\,dx$, we have

$$L_1 = |(Kf, \chi_I Tf)_{fdx}| \leq \hat{\sigma}(I,T,f) + \sum_{k=1}^{\infty} |((K-T)(\chi_k f), \chi_k Tf)_{fdx}|$$

$$+ C_\delta(\omega_\delta(K) + \omega_\delta(T)) \|f\|_{*\Omega 2} \|\chi_I Tf\|_{*\Omega 2} \quad (= \hat{\sigma}(I,T,f) + L_{11} + L_{12}, \text{ say}).$$

We have

$$L_{12} \leq C_\delta(\omega_\delta(K) + \omega_\delta(T)) \|T\|_{2,2} |I| \leq C_\delta \Lambda_1(K,T;\Omega) |I|.$$

In the same manner as in L_2,

$$L_{11} = \sum_{k=1}^{\infty} \left|\int_{I_k} f(x)(K-T)(\chi_k f)(x) \overline{T\{(\chi_k + \chi_{I_k^*-I_k} + \chi_{I_k^{*c}})f\}(x)}\, dx\right|$$

$$\leq \sum_{k=1}^{\infty} \left|\int_{I_k} f(x)(K-T)(\chi_k f)(x) \overline{T(\chi_k f)(x)}\, dx\right| + C_\delta \Lambda_1(K,T;\Omega) |I|$$

$$\leq (\|K\|_{2,2} + \|T\|_{2,2}) \tilde{\sigma}(T;\Omega)^{1/2} |I| + C_\delta \Lambda_1(K,T;\Omega) |I|$$

$$\leq C_\delta \Lambda_1(K,T;\Omega) |I|.$$

Thus the required inequality holds. Q.E.D.

Lemma 2.9. Let I, Ω, $K(x,y)$ and $T(x,y)$ be the same as in Lemma 2.8. For $E \subset \mathbb{R}$, we put $K_E = M_E K M_E$, where M_E is a multiplier: $g \to \chi_E g$. We define inductively $K_E^{(k)} = K_E K_E^{(k-1)}$ ($k \geq 1$; $K_E^{(0)}$ is the identity operator). We define $T_E^{(k)}$ in the same manner as $K_E^{(k)}$. Then, for any $f \in L^\infty_{\mathrm{real},1}$,

$$\tilde{\sigma}(I,K_I^{(2)},f) \leq \tilde{\sigma}(I, T_I^{(2)}, f) + \sum_{k=1}^{\infty} \tilde{\sigma}(I_k, K_{I_k}^{(2)}, f) + C_\delta \Lambda_3(K,T;\Omega),$$

where

$$\Lambda_3(K,T;\Omega) = \{\tilde{\sigma}_0(K)^{1/2} + \tilde{\sigma}_0(T)^{1/2} + \tilde{\sigma}(T;\Omega)^{1/2} + \omega_\delta(K) + \omega_\delta(T)\}$$

$$\times \{\sigma(K) + \sigma(T) + \omega_\delta(K) + \omega_\delta(T)\}^3.$$

Proof. We divide the proof into several steps.

(<u>First Step</u>). We begin by showing that

$$(2.22) \qquad \|X_I^{(j)} f\|_{*\Omega 2} \leq C_\delta(\sigma(X) + \omega_\delta(X))^j |I| \qquad (1 \leq j \leq 3,\ X = K, T).$$

For any $g \in L^2$ supported on $\overline{I}$, we have, with I_k^{**} = (the double of I_k^*),

$$\int_{I_k^*} |Xg(x)|^2\,dx \leq 3\int_{I_k^*} |X(\chi_{I_k^{**}} g)(x)|^2\,dx$$

$$+ 3\int_{I_k^*} |X(\chi_{I_k^{**c}} g)(x) - (X(\chi_{I_k^{**c}} g))_{I_k}|^2\,dx$$

$$+ 3\,|I_k^*|\,|(X(\chi_{I_k^{**c}} g))_{I_k}|^2 \leq 3\,\|X\|_{2,2}^2\,\|\chi_{I_k^{**}} g\|_2^2$$

$$+ C_\delta\,\omega_\delta(X)^2\,\|\chi_k Mg\|_2^2 + 6\,|I_k|\,|(X(\chi_{I_k^{**c}} g))_{I_k}|^2$$

and

$$|I_k|\,|(X(\chi_{I_k^{**c}} g))_{I_k}|^2 \leq 2|I_k|\,|(Xg)_{I_k}|^2 + 2|I_k|\,|(X(\chi_{I_k^{**}} g))_{I_k}|^2$$

$$\leq 2\,|I_k|\,|(Xg)_{I_k}|^2 + C_\delta(\sigma(K) + \omega_\delta(K))^2\,\|\chi_{I_k^{**}} g\|_2^2.$$

Thus

$$\sum_{k=1}^\infty \int_{I_k^*} |Xg(x)|^2\,dx \leq \text{Const} \sum_{k=1}^\infty |I_k|\,|(Xg)_{I_k}|^2$$

$$+ C_\delta(\sigma(K) + \omega_\delta(K))^2 \sum_{k=1}^\infty (\|\chi_k Mg\|_2^2 + \|\chi_{I_k^{**}} g\|_2^2)$$

$$\leq C_\delta(\sigma(K) + \omega_\delta(K))^2\,\{\|g\|_2^2 + \sum_{k=1}^\infty \|\chi_{I_k^{**}} g\|_2^2\},$$

which shows that

$$\|g\|_{*\Omega 2} \leq C_\delta(\sigma(K) + \omega_\delta(K))\,\|g\|_{**\Omega 2},$$

where $\|\cdot\|_{**\Omega 2}$ is defined by (2.21) with I_k^* replaced by I_k^{**}. If $j = 1$, we put $g = \chi_I f$. Then this inequality gives (2.22). If $j = 2$, we put $g = X_I f$ and use the above argument. To estimate $\|X_I f\|_{**\Omega 2}$, we use again the above argument with I_k^{**} replaced by the double of I_k^{**}. Then we obtain consequently (2.22). If $j = 3$, we put $g = X_I^{(2)} f$ and use the above argument 3 times. Then we obtain (2.22).

(<u>Second Step</u>). We show that, for any $u, v \in L^2$ supported on $\bar{I}$,

$$\text{(2.23)} \quad \sum_{k=1}^\infty |(X_k u, Y_I v)| \leq \sum_{k=1}^\infty |(X_k u, Y_k v)| + C_\delta\,\{\omega_\delta(Y)(\sigma(X) + \omega_\delta(X))$$

$$+ \tilde{\sigma}_0(X)^{1/2}(\sigma(Y) + \omega_\delta(Y))\}\,\|u\|_2\,\|v\|_{*\Omega 2} \quad (X = K, T;\ Y = K, T),$$

where $X_k = X_{I_k}$, $Y_k = Y_{I_k}$.

We have

$$\sum_{k=1}^{\infty} |(X_k u, Y_I v)| \leqq \sum_{k=1}^{\infty} |(X_k u, Y_k v)| + \sum_{k=1}^{\infty} |(X_k u, Y(\chi_{I_k^*-I_k} v))|$$

$$+ \sum_{k=1}^{\infty} |(X_k u, Y(\chi_{I_k^{*c}} v))| \quad (= \sum_{k=1}^{\infty} |(X_k u, Y_k v)| + L_1 + L_2, \text{ say}).$$

Put $\xi_k = (X_k u)_{I_k}$ $(k \geqq 1)$. Then

$$|\xi_k| = |(uX_k \chi_k)_{I_k}| \leqq \tilde{\sigma}_0(X)^{1/2} \left\{ \frac{1}{|I_k|} \int_{I_k} |u(y)|^2 \, dy\right\}^{1/2} \quad (k \geqq 1).$$

We have, with x_k' = (the midpoint of I_k),

$$L_2 = \sum_{k=1}^{\infty} \Big| \int_{I_k} (X_k u(x) - \xi_k) \left\{\int_{I_k^{*c}} (Y(x,y) - Y(x_k',y))v(y)dy\right\} dx$$

$$+ \xi_k \int_{I_k} Y(\chi_{I_k^{*c}} v)(x)\, dx \Big|$$

$$\leqq C_\delta\, \omega_\delta(Y) \sum_{k=1}^{\infty} (\|\chi_k(X_k u)\, Mv\|_1 + |\xi_k|\, \|\chi_k\, Mv\|_1)$$

$$+ \sum_{k=1}^{\infty} |\xi_k| (\|\chi_k\, Yv\|_1 + \|\chi_k Y(\chi_{I_k^*-I_k} v)\|_1)$$

$$\leqq C_\delta\, \omega_\delta(Y)\, (\|X\|_{2,2} + \tilde{\sigma}_0(X)^{1/2})\, \|u\|_2\, \|v\|_2$$

$$+ C_\delta\, \tilde{\sigma}_0(X)^{1/2}\, (\|Y\|_{2,2} + \omega_\delta(Y))\, \|u\|_2\, \|v\|_{*\Omega 2}$$

$$\leqq C_\delta \{ \omega_\delta(Y)(\sigma(X) + \omega_\delta(X)) + \tilde{\sigma}_0(X)^{1/2}(\sigma(Y) + \omega_\delta(Y))\}\, \|u\|_2\, \|v\|_{*\Omega 2}$$

and

$$L_1 \leqq \omega_\delta(Y) \sum_{k=1}^{\infty} \int_{I_k} |X_k u(x)| \left(\int_{I_k^*-I_k} \frac{|v(y)|}{|x-y|} dy\right) dx$$

$$= \text{Const}\ \omega_\delta(Y) \sum_{k=1}^{\infty} \int_{I_k^*-I_k} |v(y)|\,|H|X_k u|(y)|\,dy$$

$$\leqq C_\delta\, \omega_\delta(Y)(\sigma(X) + \omega_\delta(X))\, \|u\|_2\, \|v\|_{*\Omega 2}.$$

Hence (2.23) holds.

(Third Step). We show that

$$(2.24) \qquad \sum_{k=1}^{\infty} |(K_k f, K_I^{(3)} f)| \leqq \sum_{k=1}^{\infty} \tilde{\sigma}(I_k, K_k^{(2)}, f) + C_\delta\, \Lambda_3(K)\, |I|,$$

where

$$\Lambda_3(K) = (\tilde{\sigma}_0(K)^{1/2} + \omega_\delta(K))(\sigma(K) + \omega_\delta(K))^3.$$

Using (2.23) with $u = \chi_I f$, $v = K_I^{(2)} f$, $X = Y = K$ and using (2.22), we have

$$\sum_{k=1}^{\infty} |(K_k f, K_I^{(3)} f)| \leq \sum_{k=1}^{\infty} |(K_k f, K_k K_I^{(2)} f)|$$

$$+ C_\delta (\tilde{\sigma}_0(K)^{1/2} + \omega_\delta(K))(\sigma(K) + \omega_\delta(K)) \|\chi_I f\|_2 \|K_I^{(2)} f\|_{*\Omega 2}$$

$$\leq \sum_{k=1}^{\infty} |(K_k^{(2)} f, K_I^{(2)} f)| + C_\delta \Lambda_3(K) |I|.$$

Using (2.23) with $u = \Sigma_{k=1}^{\infty} K_k f$, $v = K_I f$, $X = Y = K$, we have

$$\sum_{k=1}^{\infty} |(K_k^{(2)} f, K_I^{(2)} f)| \leq \sum_{k=1}^{\infty} |(K_k^{(3)} f, K_I f)| + C_\delta \Lambda_3(K) |I|.$$

Using (2.23) with $u = \Sigma_{k=1}^{\infty} K_k^{(2)} f$, $v = \chi_I f$, $X = Y = K$, we have

$$\sum_{k=1}^{\infty} |(K_k^{(3)} f, K_I f)| \leq \sum_{k=1}^{\infty} \tilde{\sigma}(I_k, K_k^{(2)}, f) + C_\delta \Lambda_3(K) |I|.$$

Thus (2.24) holds.

(<u>Fourth Step</u>). We show that

$$(2.25) \quad \sum_{k=1}^{\infty} |(Z_k T_I^{(j)} f, K_I^{(3-j)} f)| \leq C_\delta \Lambda_3(K,T;\Omega)|I| \quad (1 \leq j \leq 3,\ Z = K, T).$$

If $j = 1$, we use (2.23) with $u = K_I^{(2)} f$, $v = \chi_I f$, $X = Z$, $Y = T$. Then

$$\sum_{k=1}^{\infty} |(Z_k T_I f, K_I^{(2)} f)| = \sum_{k=1}^{\infty} |(Z_k K_I^{(2)} f, T_I f)| \leq \sum_{k=1}^{\infty} |(Z_k K_I^{(2)} f, T_k f)|$$

$$+ C_\delta \{\omega_\delta(T)(\sigma(Z) + \omega_\delta(Z)) + \tilde{\sigma}_0(Z)^{1/2}(\sigma(T) + \omega_\delta(T))\} \|K_I^{(2)} f\|_2 \|f\|_{*\Omega 2}$$

$$\leq C_\delta \tilde{\sigma}(T;\Omega)^{1/2}(\sigma(Z) + \omega_\delta(Z))(\sigma(K) + \omega_\delta(K))^2 |I|$$
$$+ C_\delta \{\omega_\delta(T)(\sigma(Z) + \omega_\delta(Z)) + \tilde{\sigma}_0(Z)^{1/2}(\sigma(T) + \omega_\delta(T))\} (\sigma(K) + \omega_\delta(K))^2 |I|$$

$$\leq C_\delta \Lambda_3(K, T; \Omega) |I|.$$

If $j = 2$, we use (2.23) with $u = K_I f$, $v = T_I f$, $X = Z$, $Y = T$. Then

$$\sum_{k=1}^{\infty} |(Z_k T_I^{(2)} f, K_I f)| = \sum_{k=1}^{\infty} |(Z_k K_I f, T_I^{(2)} f)|$$

$$\leq \sum_{k=1}^{\infty} |(Z_k K_I f, T_k T_I f)| + C_\delta \Lambda_3(K, T; \Omega) |I|.$$

Using (2.23) with $u = \Sigma_{k=1}^{\infty} Z_k K_I f$, $v = f$, $X = Y = T$, we have

$$\sum_{k=1}^{\infty} |(Z_k K_I f, T_k T_I f)| = \sum_{k=1}^{\infty} |(T_k Z_k K_I f, T_I f)|$$

$$\leq \sum_{k=1}^{\infty} |(T_k Z_k K_I f, T_k f)| + C_\delta \Lambda_3(K,T;\Omega) |I| \leq C_\delta \Lambda_3(K,T;\Omega) |I|.$$

If $j = 3$, we use (2.23) 3 times. Then, in the same manner, we obtain (2.25).

(<u>Final Step</u>). We now show the required inequality in our lemma. Without loss of generality, we may assume that $\mathrm{supp}(f) \subset \bar{I}$. Using Lemma 2.6 with $u = f$, $v = K_I^{(3)} f$, $d\mu = dx$ and using (2.22), (2.24), we have

$$(2.26) \quad \tilde{\sigma}(I, K_I^{(2)}, f) = |(Kf, K_I^{(3)} f)| \leq |(Tf, K_I^{(3)} f)|$$

$$+ \sum_{k=1}^{\infty} |((K_k - T_k) f, K_I^{(3)} f)| + C_\delta (\omega_\delta(K) + \omega_\delta(T)) \|f\|_{*\Omega 2} \|K_I^{(3)} f\|_{*\Omega 2}$$

$$\leq |(T_I f, K_I^{(3)} f)| + \sum_{k=1}^{\infty} |(K_k f, K_I^{(3)} f)| + C_\delta \Lambda_3(K,T;\Omega) |I|$$

$$\leq |(T_I f, K_I^{(3)} f)| + \sum_{k=1}^{\infty} \tilde{\sigma}(I_k, K_k^{(2)}, f) + C_\delta \Lambda_3(K,T;\Omega) |I|.$$

Using Lemma 2.6 with $u = T_I f$, $v = K_I^{(2)} f$ and using (2.22), (2.25), we have

$$|(T_I f, K_I^{(3)} f)| = |(K_I T_I f, K_I^{(2)} f)| \leq |(T_I^{(2)} f, K_I^{(2)} f)|$$

$$+ \sum_{k=1}^{\infty} |((K_k - T_k) T_I f, K_I^{(2)} f)| + C_\delta \Lambda_3(K,T;\Omega) |I|$$

$$\leq |(T_I^{(2)} f, K_I^{(2)} f)| + C_\delta \Lambda_3(K,T;\Omega) |I|.$$

Repeating this argument 2 times, we obtain

$$(2.27) \quad |(T_I^{(2)} f, K_I^{(2)} f)| \leq \tilde{\sigma}(I, T_I^{(2)}, f) + C_\delta \Lambda_3(K,T;\Omega) |I|.$$

Thus (2.26) and (2.27) give the required inequality in our lemma.

§2.5. Proof of Theorem A by perturbation ([45])

In this section, we deduce Theorem A from the boundedness of the Hilbert transform and Lemma 2.9. (We do not use Cotlar's lemma nor the Fourier transform.) Fixing $0 < \varepsilon < 1/2$, we define

$$S[a](x,y) = \lambda_\varepsilon(x-y)\, T[a](x,y),$$

where

$$\lambda_\varepsilon(s) = \begin{cases} 0 & (|s| \leqq \varepsilon,\ |s| > 1/\varepsilon) \\ (|s|/\varepsilon) - 1 & (\varepsilon < |s| \leqq 2\varepsilon) \\ 1 & (2\varepsilon < |s| \leqq 1/(2\varepsilon)) \\ 2\,(1 - \varepsilon|s|) & (1/(2\varepsilon) < |s| \leqq 1/\varepsilon). \end{cases}$$

We shall show that $\|S[a]\|_{2,2} \leqq \text{Const}\ \|a\|_\infty$. Once this is known, Fatou's lemma gives Theorem A. Without loss of generality, we may assume that $a \in L^\infty_{\text{real},1}$. Since $\omega_1(S[a]) \leqq \text{Const}$, Lemma 2.5 shows that $\|S[a]\|_{2,2} \leqq \text{Const}\{\sigma(S[a]) + 1\}$. Hence it is sufficient to show that

(2.28) $$\sigma_S = \sup\ \{\sigma(S[a]);\ a \in L^\infty_{\text{real},1}\} \leqq \text{Const}\ .$$

Let

$$\tilde\sigma_{S^{(2)}} = \sup\ \{\ \frac{1}{|I|}\ \tilde\sigma(I,\ S[a]_I^{(2)},\ f);\ a,\ f \in L^\infty_{\text{real},1}\ ,\ I \text{ interval}\},$$

where $S[a]_I^{(2)}$ is defined in the same manner as $K_I^{(2)}$ in Lemma 2.9. Then Shwartz's inequality shows that $\sigma_S^4 \leqq \tilde\sigma_{S^{(2)}}$. From the definition of $S[\cdot]$, $\tilde\sigma_{S^{(2)}}$ is finite. For a, $f \in L^\infty_{\text{real},1}$ and an interval I, we show that

(2.29) $$\frac{1}{|I|}\ \tilde\sigma(I,\ S[a]_I^{(2)},\ f)\ \leqq \{(\ \frac{2}{3}\ )^4 + \frac{3}{4}\ \}\ \tilde\sigma_{S^{(2)}} + \text{Const}\ \{\sigma_S^3 + 1\}.$$

Considering $-a$ if necessary, we may assume that $(a)_I \geqq 0$. RSL of Type 1 (-1/3 -r., 1-a.) shows that there exists $b \in L^\infty_{\text{real},1}$ such that, with

$\Omega = \{x \in I;\ A(x) \neq B(x)\}$ $(B(x) = A(x_0) + \int_{x_0}^x b(s)ds$, x_0 is the left endpoint of I),

(2.30) $$-1/3 \leqq b(x) \leqq 1 \quad \text{a.e. on } I, \quad b(x) = -1/3 \quad \text{on } \Omega,$$

(2.31) $$|\Omega| \leqq \frac{1-(b)_I}{1+(1/3)}\ |I|\ \leqq \frac{1-(a)_I}{1+(1/3)}\ |I|\ \leqq \frac{3}{4}\ |I|.$$

(The function b obtained from RSL is defined only on I. Since (2.30), (2.31) are independent of the behavior outside I, we may put $b(x) = 1$ $(x \in I^c)$.) Since $A(x) = B(x)$ $(x \in I - \Omega)$, we have $S[a](x,y) = S[b](x,y)$ $(x,\ y \in I - \Omega)$. Using Lemma 2.9 with $K = S[a]$, $T = S[b]$, $\delta = 1$, we have

$$\tilde\sigma(I,\ S[a]_I^{(2)}, f) \leqq \tilde\sigma(I,\ S[b]_I^{(2)},\ f)\ +\ \sum_{k=1}^\infty\ \tilde\sigma(I_k,\ S[a]_{I_k}^{(2)},\ f)$$

$$+ \text{Const}\ \Lambda_3(S[a],\ S[b];\ \Omega)\ |I|,$$

where $\Omega = \bigcup_{k=1}^{\infty} I_k$ $(I_k = I_{\Omega,k})$. The second quantity in the above inequality is dominated by $(3/4)\,\tilde{\sigma}_{S^{(2)}}|I|$. Put $\tilde{b} = b - (1/3)$. Then $\|\tilde{b}\|_\infty \leqq 2/3$ and $S[b] = S[\tilde{b}] + (\pi/3)H$, and hence

$$\tilde{\sigma}(I, S[b]_I^{(2)}, f) \leqq \tilde{\sigma}(I, S[\tilde{b}]_I^{(2)}, f) + \text{Const}\ \|S[\tilde{b}]\|_{2,2}^3\ |I|$$

$$\leqq (\tfrac{2}{3})^4\ \tilde{\sigma}_{S^{(2)}}\ |I| + \text{Const}\ \{\sigma_S^3 + 1\}\ |I|.$$

Thus

$$\tilde{\sigma}(I, S[a]_I^{(2)}, f) \leqq \{(\tfrac{2}{3})^4 + \tfrac{3}{4}\}\ \tilde{\sigma}_{S^{(2)}}\ |I|$$

$$+ \text{Const}\ \{\sigma_S^3 + 1\}|I| + \text{Const}\ \Lambda_3(S[a], S[b]; \Omega)\ |I|.$$

It is necessary to estimate $\Lambda_3(S[a], S[b]; \Omega)$. Integration by parts shows that, for any interval J and $x \in J$,

$$|S[a]\,\chi_J(x)| \leqq \left|\int_J \frac{\lambda_\varepsilon(x-y)}{x-y}\ a(y)dy\right| + \text{Const}.$$

Lemma 2.5 shows that

$$\int_J \left|\int_J \frac{\lambda_\varepsilon(x-y)}{x-y}\ a(y)dy\right| dx \leqq \text{Const}\ \|H^*(\chi_J a)\|_2\ |J|^{1/2}$$

$$\leqq \text{Const}\ (\sigma(H) + 1)\ |J| \leqq \text{Const}\ |J|.$$

Hence we have $(1/|J|)\ \tilde{\sigma}(J, S[a], \chi_J) \leqq \text{Const}$. Taking the supremum over all J, $\tilde{\sigma}_0(S[a]) \leqq \text{Const}$. In the same manner, $\tilde{\sigma}_0(S[b]) \leqq \text{Const}$. Since $b(x) = -1/3$ on Ω, we have, for $g \in L^\infty_{\text{real},1}$,

$$\tilde{\sigma}(I_k, S[b], g) = \text{Const}\ \tilde{\sigma}(I_k, H, g) \leqq \text{Const}\ |I_k| \quad (k \geqq 1),$$

and hence $\tilde{\sigma}(S[b]; \Omega) \leqq \text{Const}$. Consequently,

$$\Lambda_3(S[a], S[b]; \Omega) \leqq \text{Const}\ \{\sigma(S[a]) + \sigma(S[b]) + 1\}^3$$

$$\leqq \text{Const}\ (\sigma_S + 1)^3,$$

which gives (2.29).

Taking the supremum of $(1/|I|)\ \tilde{\sigma}(I, S[a]_I^{(2)}, f)$ over all $a, f \in L^\infty_{\text{real},1}$ and all intervals I, we obtain, by (2.29),

$$\tilde{\sigma}_{S^{(2)}} \leqq \{(\tfrac{2}{3})^4 + \tfrac{3}{4}\}\ \tilde{\sigma}_{S^{(2)}} + \text{Const}\ \{\sigma_S^3 + 1\}.$$

Since $(2/3)^4 + (3/4) < 1$ and $\sigma_S^4 \leq \tilde{\sigma}_{S^{(2)}}$, this inequality yields (2.28). This completes the proof of Theorem A.

§2.6. Proof of Theorem B by perturbation ([17])

In this section, we deduce Theorem B from the boundedness of H. Here are two lemmas necessary for the proof.

Lemma 2.10 (Calderón [4]). $\|T_n[a]\|_{2,2} \leq (\mathrm{Const})^n \|a\|_\infty^n \quad (n \geq 1)$.

Proof. Fixing $0 < \varepsilon < 1/2$, we put

$$S_n[a](x,y) = \lambda_\varepsilon(x-y)\, T_n[a](x,y) .$$

Since $\omega_1(S_n[a]) \leq \mathrm{Const}\, n\, \|a\|_\infty^n$, it is sufficient to show that

$$(2.32) \qquad \sigma_{S_n} = \sup \{\sigma(S_n[a]);\ a \in L^\infty_{\mathrm{real},1}\} \leq C_0^{n+1}$$

for some absolute constant C_0 (which will be determined later). Let $S_0[\cdot] = (-\pi)H$. Then $\sigma_{S_0} = \sigma(-\pi H) \leq \pi$. Suppose that $\sigma_{S_k} \leq C_0^{k+1}$ $(0 \leq k \leq n-1)$. We define $\tilde{\sigma}_{S_n^{(2)}}$ in the same manner as $\tilde{\sigma}_{S^{(2)}}$. Then $\sigma_{S_n}^4 \leq \tilde{\sigma}_{S_n^{(2)}}$. Let a, b, f, I, Ω be the same as in §2.5. Using Lemma 2.9 with $K = S_n[a]$, $T = S_n[b]$, $\delta = 1$, we have

$$(2.33) \qquad \tilde{\sigma}(I, S_n[a]_I^{(2)}, f) \leq \tilde{\sigma}(I, S_n[b]_I^{(2)}, f) + \sum_{k=1}^{\infty} \tilde{\sigma}(I_k, S_n[a]_I^{(2)}, f)$$
$$+ \mathrm{Const}\ \Lambda_3(S_n[a], S_n[b];\Omega)\ |I| .$$

Put $b^* = (3/2)(b - (1/3))$. Then $\|b^*\|_\infty \leq 1$ and

$$S_n[b] = S_n[\tfrac{2}{3} b^* + \tfrac{1}{3}] = \sum_{k=0}^{n} \binom{n}{k} \left(\tfrac{2}{3}\right)^k \left(\tfrac{1}{3}\right)^{n-k} S_k[b^*].$$

By Lemma 2.5 and the assumption of our induction, we have

$$\|S_k[d]\|_{2,2} \leq \mathrm{Const}\ \{\sigma_{S_k} + \omega_1(S_k[d])\}$$
$$\leq \mathrm{Const}\ (C_0^n + n) \qquad (d \in L^\infty_{\mathrm{real},1},\ 0 \leq k \leq n-1).$$

Hence, in the same manner as in §2.5, we obtain, by (2.33),

$$\frac{1}{|I|}\tilde{\sigma}(I,\ S_n[a]_I^{(2)},\ f) \leq \{(\tfrac{2}{3})^{4n} + \tfrac{3}{4}\}\ \tilde{\sigma}_{S_n^{(2)}}$$
$$+ \text{Const}\ (C_0^n + n)(\sigma_{S_n} + n)^3 + \text{Const}\ (C_0^n + n)^4 .$$

Since $a, f \in L^\infty_{real,1}$ and I are arbitrary, this inequality holds with the first quantity replaced by $\tilde{\sigma}_{S_n^{(2)}}$. Since $(2/3)^{4n} + (3/4) \leq 0.99$ and $\sigma^4_{S_n} \leq \tilde{\sigma}_{S_n^{(2)}}$, we have $\sigma_{S_n} \leq C_0' (C_0^n + n)$ for some absolute constant C_0'. Let $C_0 = \max\{2\,C_0', \pi\}$. Then $\sigma_{S_n} \leq C_0^{n+1}$. This shows that (2.32) holds for all $n \geq 0$. This completes the proof of Lemma 2.10. Q.E.D.

Remark 2.11. It is known that

$$\|T_n[a]\|_{\infty,BMO} \leq \text{Const}\ \{\|T_n[a]\|_{2,2} + \omega_1(T_n[a])\}$$
$$\leq \text{Const}\ \{\|T_n[a]\|_{2,2} + (n+1)\} \qquad (n \geq 0,\ \ a \in L^\infty_{real,1}),$$

where $\|T_n[a]\|_{\infty,BMO}$ is the norm of $T_n[a]$ from L^∞ to BMO. (See Lemma 2.5.) The proof of Lemma 2.10 by Theorem 1.12 is as follows. Let $a \in L^\infty_{real,1}$ and $n \geq 1$. Integration by parts shows that $T_n[a]1 = T_{n-1}[a]a$. Hence Theorem 1 shows that

$$\|T_n[a]\|_{2,2} \leq \text{Const}\ \{\|T_n[a]1\|_{BMO} + n\}$$
$$= \text{Const}\ \{\|T_{n-1}[a]a\|_{BMO} + n\} \leq \text{Const}\ \{\|T_{n-1}[a]\|_{\infty,BMO} + n\}$$
$$\leq \text{Const}\ \{\|T_{n-1}[a]\|_{2,2} + n\},$$

which gives Lemma 2.10.

For $a \in L^\infty_{real}$, we define a kernel

$$\text{(2.34)} \qquad E[a](x,y) = \frac{1}{x-y}\ \exp\{i\,\frac{A(x)-A(y)}{x-y}\},$$

where A is a primitive of a. The following lemma was first shown by Coifman-McIntosh-Meyer [7]. A proof by perturbation was given by David [17].

Lemma 2.12. There exists an absolute constant N_0 such that

$$\|E[a]\|_{2,2} \leq \text{Const}(1 + \|a\|_\infty)^{N_0} \qquad (a \in L^\infty_{real}).$$

Proof. Since $\omega_1(E[a]) \leq \text{Const}\ (1 + \|a\|_\infty)$, it is sufficient to show that

$$\sigma(E[a]) \leq \text{Const}\ (1 + \|a\|_\infty)^{N_0}$$

for some absolute constant $N_0 \geqq 1$. We put

$$(2.35) \qquad \sigma_E(\beta) = \sup \{\sigma(E[a]);\ a \in L^\infty_{real,\beta}\} \qquad (\beta > 0)$$

and show that

$$(2.36) \qquad \sigma_E(\beta) \leqq \mathrm{Const}(1 + \beta)^{N_0} \qquad (\beta > 0).$$

Lemma 2.10 shows that, for any $a \in L^\infty_{real}$,

$$\sigma(E[a]) = \sigma\left(\sum_{n=0}^{\infty} \frac{i^n}{n!} T_n[a]\right) \leqq \sum_{n=0}^{\infty} \frac{1}{n!} \sigma(T_n[a])$$

$$\leqq \sum_{n=0}^{\infty} \frac{1}{n!} (\mathrm{Const})^n \|a\|_\infty^n \leqq \exp \{\mathrm{Const}(1 + \|a\|_\infty)\} .$$

Hence $\sigma_E(\beta) < \infty$ for all $\beta > 0$ and $\sigma_E(1) \leqq \mathrm{Const}$. Let $\beta > 1$, $a \in L^\infty_{real,\beta}$, $f \in L^\infty_{real,1}$ and I be an interval. RSL of Type 1 ($-\beta/3 - r.,\beta - a.$) shows that there exists $b \in L^\infty_{real}$ such that, with $\Omega = \{x \in I;\ A(x) \neq B(x)\}$,

$$-\beta/3 \leqq b(x) \leqq \beta \quad \text{a.e. on } I, \qquad b(x) = -\beta/3 \quad \text{on } \Omega ,$$

$$|\Omega| \leqq \frac{\beta - (b)_I}{\beta + (\beta/3)} |I| \leqq \frac{3}{4} |I|.$$

Using Lemma 2.7 with $K = E[a]$, $T = E[b]$, $\delta = 1$, we have

$$\sigma(I, E[a], f) \leqq \sigma(I, E[b], f) + \sum_{k=1}^{\infty} \sigma(I_k, E[a], f)$$

$$+ \mathrm{Const}\ \{\tilde{\sigma}(E[b];\Omega)^{1/2} + \omega_1(E[a]) + \omega_1(E[b])\}\ |I|,$$

where $I_k = I_{\Omega,k}$ $(k \geqq 1)$. Put $\tilde{b} = b - (\beta/3)$. Then $\|\tilde{b}\|_\infty \leqq 2\beta/3$ and $\sigma(I, E[b], f) = \sigma(I, E[\tilde{b}], f)$. We have $\omega_1(E[a]) + \omega_1(E[b]) \leqq \mathrm{Const}\ \beta$. Since $b(x) = -\beta/3$ on Ω, we have

$$\tilde{\sigma}(E[b];\ \Omega) = \pi^2 \tilde{\sigma}(H;\Omega) \leqq \mathrm{Const}.$$

Thus

$$\frac{1}{|I|} \sigma(I, E[a], f) \leqq \sigma_E\left(\frac{2\beta}{3}\right) + \frac{3}{4}\ \sigma_E(\beta) + \mathrm{Const}\ \beta ,$$

which yields that

$$\sigma_E(\beta) \leqq \sigma_E\left(\frac{2\beta}{3}\right) + \frac{3}{4}\ \sigma_E(\beta) + \mathrm{Const}\ \beta ,$$

that is,

(2.37) $\sigma_E(\beta) \leqq 4\,\sigma_E(\frac{2\beta}{3}) + \text{Const}\,\beta$.

Let n be the minimum of integers $k \geqq 1$ such that $(2/3)^k \beta \leqq 1$. Then $n \leqq (\log \beta)/(\log 3/2) + \text{Const}$. Inequality (2.37) shows that

$$\sigma_E(\beta) \leqq 4^n\,\sigma_E((\tfrac{2}{3})^n\beta) + \text{Const} \sum_{k=0}^{n-1} 4^k (\tfrac{2}{3})^k \beta$$

$$\leqq 4^n \{\sigma_E(1) + \text{Const}\,(\tfrac{2}{3})^n \beta\}$$

$$\leqq \text{Const}\ 4^n \leqq \text{Const}\,(1+\beta)^{N_0},$$

where $N_0 = (\log 4)/(\log 3/2)$. Thus (2.36) holds. This completes the proof of Lemma 2.12. Q.E.D.

We now give the proof of Theorem B. Since $1/(1+ix) = \int_0^\infty e^{-ixs} e^{-s} ds$ $(x \in \mathbb{R})$, we have

$$C[a] = \int_0^\infty E[-sa]e^{-s}\,ds.$$

By Lemma 2.12, we have

$$\|C[a]\|_{2,2} \leqq \int_0^\infty \|E[-sa]\|_{2,2}\, e^{-s}\,ds$$

$$\leqq \text{Const} \int_0^\infty (1+s\|a\|_\infty)^{N_0} e^{-s}\,ds \leqq \text{Const}\,(1+\|a\|_\infty)^{N_0}.$$

This completes the proof of Theorem B.

§2.7. Estimates of norms of E[·] and C[·]

In this section, we show

Theorem C ([44]). For any real-valued function a in BMO,

(2.38) $\|E[a]\|_{2,2} \leqq \text{Const}\,(1+\|a\|_{BMO})$,

(2.39) $\|C[a]\|_{2,2} \leqq \text{Const}\,(1+\sqrt{\|a\|_{BMO}})$.

In [7], (2.38) was given with $\|a\|_{BMO}$ replaced by $\|a\|_{BMO}^9$, and (2.39) was given with $\sqrt{\|a\|_{BMO}}$ replaced by $\|a\|_{BMO}^8$. This theorem was established in [44] via [42], [43], [50]. In this note, we deduce Theorem C from (2.36). We use RSL repeatedly. (The sun also rises!) RSL in §2.2 is called of Type 1 (γ - *r.*, β - *a.*) The lower bound of a(x) is independent of (2.5)-(2.8). Let a(x), I, α, β and A(x) be the same as in RSL in §2.2. For $0 \leqq \gamma \leqq \beta$, we define analogously B(x) by using the sun at the right upper infinity of angle arctan γ. Then

$$\alpha \leqq b(x) \leqq \gamma \qquad \text{a.e. on } I, \qquad b(x) = \gamma \qquad (x \in I_k,\ k \geqq 1),$$

$$(a)_{I_k} \geqq \gamma \qquad (k \geqq 1), \qquad |\Omega| \leqq \frac{(-\alpha) + (b)_I}{(-\alpha) + \gamma}\ |I|.$$

This RSL is called of Type 2 (γ-*ray*, α-*descent*). In this case β is independent of the above estimates. For $0 \leqq \gamma \leqq \beta$, we define $B(x) = \sup \Phi(x)$, where the supremum is taken over all functions Φ such that $\Phi \leqq A$, $\Phi' \leqq \gamma$ a.e. on I. We define b, Ω in the same manner. Then

$$\alpha \leqq b(x) \leqq \gamma \quad \text{a.e. on } I, \qquad b(x) = \gamma \qquad (x \in I_k,\ k \geqq 1),$$

$$(a)_{I_k} \geqq \gamma \quad (k \geqq 1), \qquad |\Omega| \leqq \frac{(-\alpha) + (b)_I}{(-\alpha) + \gamma}\ |I| .$$

This RSL is called of Type 3 (γ - *r*., α - *d*.) . In this case β is independent of the above estimates. For $\alpha \leqq \gamma \leqq 0$, we define $B(x)$, in the same manner as in Type 3, by using the sun at the lower right infinity of angle - arctan $|\gamma|$. Then

$$\gamma \leqq b(x) \leqq \beta \qquad \text{a.e. on } I, \qquad b(x) = \gamma \qquad (x \in I_k,\ k \geqq 1),$$

$$(a)_{I_k} \leqq \gamma \qquad (k \geqq 1), \quad |\Omega| \leqq \frac{\beta - (b)_I}{\beta - \gamma}\ |I|.$$

This RSL is called of Type 4 (γ - *r*., β - *a*.). In this case, α is independent of the above estimates.

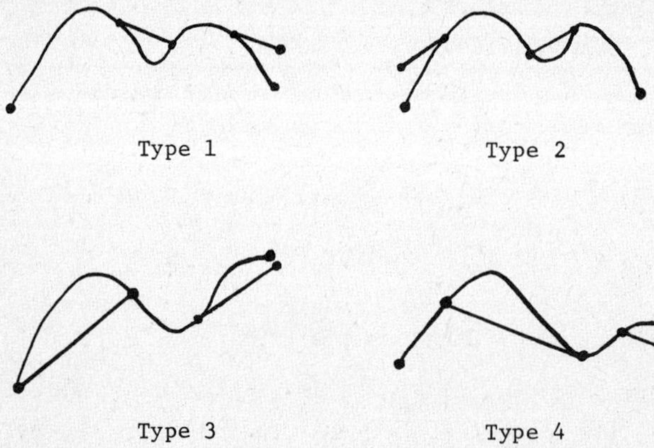

RSL of Type j is reduced to RSL of Type 1 by a suitable affine transformation.

§2.8. Proof of (2.38)

(First Step). We divide the proof into three steps. Recall $\sigma_E(\beta)$ defined by (2.35). Put

$$\hat{\sigma}_E(\beta) = \sup\{\hat{\sigma}(E[a]); a \in L^\infty_{real,\beta}\} \qquad (\beta > 0).$$

(See (2.16).) In this step, we show that, for $\beta \geqq 1$, $0 < \delta \leqq 1$,

$$\text{(2.40)} \quad \hat{\sigma}_E(\beta) \leqq \hat{\sigma}_E(\theta\beta) + \frac{1+\theta}{2\theta}\ \hat{\sigma}_E(\frac{1+\theta}{2}\beta) + (C_\delta/\theta)\ \beta^\delta\{\sigma_E(\beta) + \beta^\delta\},$$

where θ $(0 < \theta < 1)$ is determined later. For $a \in L^\infty_{real,\beta}$, $f \in L^\infty_{real}$, $0 \leqq f \leqq 1$ and an interval I, we study an a-priori estimate with respect to $\hat{\sigma}(I, E[a], f)$. Since $\hat{\sigma}(I,E[a],f) = \hat{\sigma}(I,E[-a],f)$, we may assume that $(a)_I \geqq 0$. RSL of Type 1 $(-\theta\beta - \kappa.,\ \beta - \alpha.)$ shows that there exists $b \in L^\infty_{real}$ such that, with $\Omega = \{x \in I;\ A(x) \neq B(x)\} = \cup^\infty_{k=1} I_k$ $(I_k = I_{\Omega,k})$.

$$-\theta\beta \leqq b(x) \leqq \beta \quad \text{a.e. on } I, \quad b(x) = -\theta\beta \quad (x \in I_k,\ k \geqq 1),$$

$$(a)_{I_k} \leqq -\theta\beta \quad (k \geqq 1), \qquad |\Omega| \leqq \frac{\beta - \eta}{\beta + \theta\beta}|I| \qquad (\eta = (b)_I).$$

Using Lemma 2.8 with $K = E[a]$, $T = E[b]$, we have

$$\hat{\sigma}(I, E[a], f) \leqq \hat{\sigma}(I, E[b], f) + \sum_{k=1}^{\infty} \hat{\sigma}(I_k, E[a], f) + C_\delta\ \Lambda_1(E[a], E[b]; \Omega)\ |I|.$$

Since $\omega_\delta(E[a]) \leqq C_\delta\ \beta^\delta$, $\omega_\delta(E[b]) \leqq C_\delta\ \beta^\delta$ and

$$\tilde{\sigma}(E[b];\ \Omega)^{1/2} = \text{Const}\ \tilde{\sigma}(H;\Omega)^{1/2} \leqq \text{Const},$$

we have

$$\begin{aligned} \Lambda_1(E[a],E[b];\Omega) &= \{\tilde{\sigma}(E[b];\Omega)^{1/2} + \omega_\delta(E[a]) + \omega_\delta(E[b])\} \\ &\times \{\sigma(E[a]) + \sigma(E[b]) + \omega_\delta(E[a]) + \omega_\delta(E[b])\} \\ &\leqq C_\delta\ \beta^\delta\ \{\sigma_E(\beta) + \beta^\delta\}, \end{aligned}$$

and hence

$$\hat{\sigma}(I,E[a],f) \leqq \hat{\sigma}(I,E[b],f) + \sum_{k=1}^{\infty} \hat{\sigma}(I_k,E[a],f) + C_\delta\ \beta^\delta\{\sigma_E(\beta) + \beta^\delta\}\ |I|.$$

For each I_k, we use RSL of Type 2 ($\theta\beta - r., -\beta - d.$). Since $(a)_{I_k} \leqq -\theta\beta$, there exist $b_k \in L^\infty_{real}$ and an open set $\Omega_k = \cup_{\ell=1}^\infty I_{k,\ell}$ in I_k such that

$$-\beta \leqq b_k(x) \leqq \theta\beta \quad \text{a.e. on } I_k, \quad b_k(x) = \theta\beta \quad (x \in \Omega_k),$$

$$(a)_{I_{k,\ell}} \geqq \theta\beta \quad (\ell \geqq 1),$$

$$|\Omega_k| \leqq \frac{-(-\beta)+(b_k)_{I_k}}{-(-\beta) + \theta\beta} |I_k| \leqq \frac{\beta + (a)_{I_k}}{\beta + \theta\beta} |I_k| \leqq \frac{1-\theta}{1+\theta} |I_k|.$$

Let $\tilde{b}_k = b_k - ((1-\theta)\beta/2)$. Then $\|\tilde{b}_k\|_\infty \leqq (1+\theta)\beta/2$ and $\hat{\sigma}(I_k, E[\tilde{b}_k], f) = \hat{\sigma}(I_k, E[b_k], f)$. Hence, by Lemma 2.8, we have

$$\hat{\sigma}(I_k, E[a], f) \leqq \hat{\sigma}(I_k, E[b_k], f) + \sum_{\ell=1}^\infty \hat{\sigma}(I_{k,\ell}, E[a], f)$$

$$+ C_\delta \ \Lambda_1(E[a], E[b_k]; \Omega_k) |I_k| \leqq \hat{\sigma}_E(\tfrac{1+\theta}{2}\beta) |I_k|$$

$$+ \sum_{\ell=1}^\infty \hat{\sigma}(I_{k,\ell}, E[a], f) + C_\delta \beta^\delta \{\sigma_E(\beta) + \beta^\delta\} |I_k|,$$

which yields that

$$\hat{\sigma}(I, E[a], f) \leqq \hat{\sigma}(I, E[b], f) + \hat{\sigma}_E(\tfrac{1+\theta}{2}\beta) \sum_{k=1}^\infty |I_k|$$

$$+ \sum_{k=1}^\infty \sum_{\ell=1}^\infty \hat{\sigma}(I_{k,\ell}, E[a], f) + C_\delta \beta^\delta \{\sigma_E(\beta) + \beta^\delta\}\{|I| + \sum_{k=1}^\infty |I_k|\}.$$

For each $I_{k,\ell}$, we use RSL of Type 1 ($-\theta\beta - r., \beta - a.$) Since $(a)_{I_{k,\ell}} \geqq \theta\beta$, we obtain an open set $\Omega_{k,\ell} = \cup_{m=1}^\infty I_{k,\ell,m}$ in $I_{k,\ell}$ such that

$$(a)_{I_{k,\ell,m}} \leqq -\theta\beta, \quad \sum_{m=1}^\infty |I_{k,\ell,m}| \leqq \frac{1-\theta}{1+\theta} |I_{k,\ell}| .$$

In the same manner as above,

$$\hat{\sigma}(I, E[a], f) \leqq \hat{\sigma}(I, E[b], f) + \hat{\sigma}_E(\tfrac{1+\theta}{2}\beta) \{\sum_{k=1}^\infty |I_k| + \sum_{k=1}^\infty \sum_{\ell=1}^\infty |I_{k,\ell}|\}$$

$$+ \sum_{k=1}^\infty \sum_{\ell=1}^\infty \sum_{m=1}^\infty \hat{\sigma}(I_{k,\ell,m}, E[a], f)$$

$$+ C_\delta \beta^\delta \{\sigma_E(\beta) + \beta^\delta\} \{|I| + \sum_{k=1}^\infty |I_k| + \sum_{k=1}^\infty \sum_{\ell=1}^\infty |I_{k,\ell}|\} .$$

Since $(a)_{I_{k,\ell,m}} \leqq -\theta\beta$, $\hat{\sigma}(I_{k,\ell,m}, E[a], f)$ is estimated in the same manner as in $\hat{\sigma}(I_k, E[a], f)$. Repeating this argument,

$$\hat{\sigma}(I,E[a],f) \leqq \hat{\sigma}(I,E[b],f) + \hat{\sigma}_E(\tfrac{1+\theta}{2}\beta)\ |\Omega|\ \{1 + \tfrac{1-\theta}{1+\theta} + (\tfrac{1-\theta}{1+\theta})^2 + \ldots\}$$

$$+ C_\delta\ \beta^\delta\ \{\sigma_E(\beta) + \beta^\delta\}\ \{|I| + |\Omega|\ (1 + \tfrac{1-\theta}{1+\theta} + (\tfrac{1-\theta}{1+\theta})^2 + \ldots)\}$$

$$\leqq \hat{\sigma}(I,E[b],f) + \hat{\sigma}_E(\tfrac{1+\theta}{2}\beta)\ \frac{\beta-\eta}{\beta+\theta\beta}\ \frac{1+\theta}{2\theta}\ |I| + (C_\delta/\theta)\beta^\delta\{\sigma_E(\beta) + \beta^\delta\}\ |I|$$

$$= \hat{\sigma}(I,E[b],f) + \hat{\sigma}_E(\tfrac{1+\theta}{2}\beta)\ \frac{\beta-\eta}{2\theta\beta}\ |I| + (C_\delta/\theta)\ \beta^\delta\{\sigma_E(\beta) + \beta^\delta\}\ |I|.$$

To estimate $\hat{\sigma}(I,E[b],f)$, we use RSL of Type 3 $(\theta\beta - r., -\theta\beta - d.)$ to $b(x)$ and I. There exist $c \in L^\infty_{real}$ and an open set $\Omega' = \bigcup_{j=1}^\infty I_j'$ in I such that

$$-\theta\beta \leqq c(x) \leqq \theta\beta \qquad \text{a.e. on } I,$$

$$|\Omega'| \leqq \frac{-(-\theta\beta) + (c)_I}{-(-\theta\beta) + \theta\beta}\ |I| \leqq \frac{\theta\beta+\eta}{2\theta\beta}\ |I|.$$

In the same manner as above,

$$\hat{\sigma}(I,E[b],f) \leqq \hat{\sigma}(I,E[c],f) + \sum_{j=1}^\infty \hat{\sigma}(I_j', E[b], f) + C_\delta\ \beta^\delta\ \{\sigma_E(\beta) + \beta^\delta\}\ |I|$$

$$\leqq \hat{\sigma}_E(\theta\beta)|I| + \hat{\sigma}_E(\tfrac{1+\theta}{2}\beta)\ \frac{\theta\beta+\eta}{2\theta\beta}\ |I| + C_\delta\ \beta^\delta\ \{\sigma_E(\beta) + \beta^\delta\}\ |I|.$$

Consequently,

$$\frac{1}{|I|}\ \hat{\sigma}(I,E[a],f) \leqq \{\hat{\sigma}_E(\theta\beta) + \hat{\sigma}_E\ (\tfrac{1+\theta}{2}\beta)\ \frac{\theta\beta+\eta}{2\theta\beta} + C_\delta\ \beta^\delta(\sigma_E(\beta) + \beta^\delta)\}$$

$$+ \{\hat{\sigma}_E(\tfrac{1+\theta}{2}\beta)\ \frac{\beta-\eta}{2\theta\beta} + (C_\delta/\theta)\beta^\delta(\sigma_E(\beta) + \beta^\delta)\}$$

$$\leqq \hat{\sigma}_E(\theta\beta) + \frac{1+\theta}{2\theta}\ \hat{\sigma}_E(\tfrac{1+\theta}{2}\beta) + (C_\delta/\theta)\ \beta^\delta\{\sigma_E(\beta) + \beta^\delta\},$$

which gives (2.40).

(<u>Second Step</u>). In this step, from (2.40), we deduce

$$(2.41) \qquad \sigma_E(\beta) \leqq \text{Const}\ \beta \qquad (\beta \geqq 1).$$

Let

$$h_\lambda(\theta) = \theta^\lambda + \frac{1+\theta}{2\theta}\ (\frac{1+\theta}{2})^\lambda \qquad (0 < \theta < 1,\ \lambda > 0).$$

Then $h_\lambda(\theta) > 1$ for any $0 < \theta < 1$, $\lambda > 2$ and

$$1 = h_2(1/3) = \min_{0<\theta<1} h_2(\theta).$$

Hence we choose $\theta = 1/3$ in (2.40). For any $a \in L^\infty_{real,\beta}$, $f \in L^\infty_{real,1}$ and an interval I, we have

$$\sigma(I,E[a],f) \leqq 3\,\sigma(I,E[a], \frac{f + 2\chi_I}{3}) + 3\,\sigma(I,E[a], \frac{2}{3}\chi_I)$$

$$\leqq \text{Const}\ \{\hat{\sigma}(I,E[a], \frac{f + 2\chi_I}{3})^{1/2} + \hat{\sigma}(I, E[a], \frac{2}{3}\chi_I)^{1/2}\}\ |I|^{1/2}$$

$$\leqq \text{Const}\ \hat{\sigma}_E(\beta)^{1/2}|I|,$$

and hence $\sigma_E(\beta) \leqq \text{Const}\ \hat{\sigma}_E(\beta)^{1/2}$. Inequality (2.40), $\theta = 1/3$ shows that

$$\text{(2.42)}\quad \hat{\sigma}_E(\beta) \leqq \hat{\sigma}_E(\tfrac{\beta}{3}) + 2\,\hat{\sigma}_E(\tfrac{2\beta}{3}) + C_\delta\,\beta^\delta\,\{\hat{\sigma}_E(\beta)^{1/2} + \beta^\delta\} \qquad (\beta \geqq 1).$$

By (2.36),

$$\text{(2.43)}\quad \hat{\sigma}_E(\beta) \leqq \text{Const}\ \{\sigma_E(\beta) + \beta\}^2 \leqq \text{Const}\ \beta^N \qquad (\beta \geqq 1),$$

where $N = 2N_0 + 2$. Suppose that $N \geqq 3$. We put

$$\tau_m = \sup\{\ \hat{\sigma}_E(\beta)\ \beta^{-\frac{N+1}{2}}\ ;\ 1 \leqq \beta \leqq (\tfrac{3}{2})^m\ \} \qquad (m = 3,4,\ldots).$$

Then $\tau_3 \leqq \text{Const}$. For any $m \geqq 4$ and $(3/2)^{m-1} < \beta \leqq (3/2)^m$, we have, by (2.42),

$$\hat{\sigma}_E(\beta)\ \beta^{-\frac{N+1}{2}} \leqq \beta^{-\frac{N+1}{2}}\ \{\hat{\sigma}_E(\tfrac{\beta}{3}) + 2\,\hat{\sigma}_E(\tfrac{2\beta}{3}) + C_\delta(\beta^{\frac{N}{2}+\delta} + \beta^{2\delta})\}$$

$$\leqq \beta^{-\frac{N+1}{2}}\ \{(\tfrac{\beta}{3})^{\frac{N+1}{2}}\tau_{m-1} + 2(\tfrac{2\beta}{3})^{\frac{N+1}{2}}\tau_{m-1} + C_\delta(\beta^{\frac{N}{2}+\delta} + \beta^{2\delta})\ \}$$

$$\leqq \{(\tfrac{1}{3})^{\frac{N+1}{2}} + 2(\tfrac{2}{3})^{\frac{N+1}{2}}\}\ \tau_{m-1} + C_\delta(\beta^{\delta-(1/2)} + \beta^{2\delta - \frac{N+1}{2}})\}$$

$$\leqq \tau_{m-1} + C_\delta\ \beta^{\delta-(1/2)} \leqq \tau_{m-1} + C_\delta(\tfrac{2}{3})^{(m-1)((1/2)-\delta)}$$

Hence $\tau_m \leqq \tau_{m-1} + C_\delta\ (2/3)^{(m-1)((1/2)-\delta)}$. We choose $\delta = 1/4$. Then

$$\tau_m \leqq \tau_3 + \text{Const}\ \sum_{k=3}^{m-1} (\tfrac{2}{3})^{k/4} \leqq \text{Const},$$

which gives that

$$\hat{\sigma}_E(\beta) \leqq \text{Const} \quad \beta^{\frac{N+1}{2}} \qquad (\beta \geqq 1),$$

i.e., (2.43) holds with N replaced by $(N+1)/2$. Repeating this argument, we obtain

$$\hat{\sigma}_E(\beta) \leqq \text{Const}\ \beta^3 \qquad (\beta \geqq 1).$$

Put

$$\tau'_m = \sup\ \{\hat{\sigma}_E(\beta)\ \beta^{-2};\ 1 \leqq \beta \leqq (\tfrac{3}{2})^m\} \qquad (m = 3,4,\ldots).$$

Then $\tau'_3 \leqq$ Const. Since $(1/3)^2 + 2(2/3)^2 = 1$, we have, in the same manner as above,

$$\tau'_m \leqq \tau'_{m-1} + C_\delta \quad \{(\tfrac{2}{3})^{(m-1)((1/2)-\delta)} + (\tfrac{2}{3})^{(m-1)(2-\delta)}\}$$

$$\leqq \tau'_{m-1} + C_\delta \quad (\tfrac{2}{3})^{(m-1)((1/2)-\delta)} \qquad (m \geqq 4).$$

We put $\delta = 1/4$. Then $\tau'_m \leqq$ Const, which yields that

$$\hat{\sigma}_E(\beta) \leqq \text{Const}\ \beta^2 \qquad (\beta \geqq 1).$$

Since $\sigma_E(\beta) \leqq \text{Const}\ \hat{\sigma}_E(\beta)^{1/2}$, this shows (2.41).

(Final Step). At last, we deduce (2.38) from (2.41). Let $a \in L^\infty_{real}$. We show that

$$\text{(2.44)} \qquad \omega_\delta(E[a]) \leqq C_\delta\ (1 + \beta^{2\delta}) \qquad (\beta = \|a\|_{BMO},\ 0 < \delta \leqq 1/2)$$

We have evidently $|E[a](x,y)| \leqq 1/|x-y|$. Let $x, x', y \in \mathbb{R}$ satisfy $|x-x'| \leqq |x-y|/2$. We may assume that $y < x < x'$. We have

$$\int_y^x |a(s) - (a)_{(x,x')}|\ ds \leqq \text{Const}\ \ \beta(x-y) \log \frac{x-y}{x'-y},$$

which gives that

$$\left| \frac{A(x)-A(y)}{x-y} - \frac{A(x') - A(y)}{x'- y} \right| = \left| \frac{x' - x}{(x-y)(x'-y)} \int_y^x (a(s) - (a)_{(x,x')})ds \right.$$

$$\left. - \frac{1}{x'-y} \int_x^{x'} (a(s) - (a)_{(x,x')})ds \right|$$

$$\leq \text{Const } \beta \frac{x'-x}{x-y} \{1 + \log \frac{x-y}{x'-x}\} \leq \text{Const } \beta (x'-x)^{1/2}/(x-y)^{1/2}.$$

Thus

$$|E[a](x,y) - E[a](x',y)| \leq \frac{x' - x}{(x-y)(x'-y)}$$

$$+ \frac{1}{x'-y} \left| \exp\{ i \frac{A(x)-A(y)}{x-y} \} - \exp\{ i \frac{A(x')-A(y)}{x'-y} \} \right|$$

$$\leq \frac{x' - x}{(x-y)(x'-y)} + C_\delta \frac{1}{x'-y} \left| \exp\{ i \frac{A(x)-A(y)}{x-y} \} - \exp\{ i \frac{A(x')-A(y)}{x'-y} \} \right|^{2\delta}$$

$$\leq \frac{x' - x}{(x-y)(x'-y)} + C_\delta \frac{1}{x'-y} \left| \frac{A(x)-A(y)}{x-y} - \frac{A(x')-A(y)}{x'-y} \right|^{2\delta}$$

$$\leq \text{Const } \frac{x' - x}{(x-y)^2} + C_\delta \beta^{2\delta} \frac{(x' - x)^\delta}{(x-y)^{1+\delta}} \leq C_\delta (1 + \beta^{2\delta})(x'-x)^\delta/(x-y)^{1+\delta} .$$

For $f \in L^\infty_{real,1}$ and an interval I, we estimate $\sigma(I,E[a],f)$. Let $\tilde{a} = a - (a)_I$. Then $\sigma(I,E[a],f) = \sigma(I,E[\tilde{a}],f)$. Lemma 2.2 shows that there exists an open set $\Omega = \cup_{k=1}^\infty I_k$ $(I_k = I_{\Omega,k})$ in I such that

$$|\Omega| \leq \frac{1}{2\beta} \int_I |\tilde{a}(s)| ds \leq |I|/2, \quad (|\tilde{a}|)_{I_k} \leq 2\beta \qquad (k \geq 1),$$

$$|\tilde{a}(x)| \leq 2\beta \qquad \text{a.e. on } I - \Omega .$$

We put

$$\tilde{b}(x) = \begin{cases} \tilde{a}(x) & (x \in I - \Omega) \\ (\tilde{a})_{I_k} & (x \in I_k,\ k \geq 1) \\ 0 & (x \notin \Omega). \end{cases}$$

Then $\tilde{b} \in L^\infty_{real,2\beta}$. Using Lemma 2.7 with $K = E[\tilde{a}]$, $T = E[\tilde{b}]$, we have

$$\sigma(I,E[a],f) = \sigma(I,E[\tilde{a}],f)$$

$$\leq \sigma(I,E[\tilde{b}],f) + \sum_{k=1}^\infty \sigma(I_k,E[\tilde{a}],f) + C_\delta (1 + \beta^{2\delta}) |I|$$

$$\leq \sigma_E(2\beta) |I| + \sum_{k=1}^\infty \sigma(I_k,E[\tilde{a}],f) + C_\delta (1 + \beta^{2\delta}) |I|.$$

For each I_k, we use Lemma 2.2 with $\tilde{a} - (\tilde{a})_{I_k}$. Repeating this discussion, we have

$$\sigma(I, E[a], f) \leqq \sigma_E(2\beta)\ |I|\ \{1 + \frac{1}{2} + \frac{1}{4} + \dots \}$$

$$+ C_\delta\ (1 + \beta^{2\delta})\ |I|\ \ \{1 + \frac{1}{2} + \frac{1}{4} + \dots \} \leqq \ \{2\,\sigma_E(2\beta) + C_\delta\ (1 + \beta^{2\delta})\ \}\ |I|\ .$$

Thus

$$\sigma(E[a]) \leqq 2\,\sigma_E(2\beta) + C_\delta\ (1 + \beta^{2\delta})$$

$$\leqq \text{Const}\ (1 + \beta) = \text{Const}(1 + \|a\|_{BMO}).$$

Consequently, Lemma 2.5 and (2.44) yield (2.38) in the case where $a \in L^\infty_{real}$. Let a be a real-valued function in BMO. We put

$$a_n(x) = \left\{\begin{array}{lll} n & \text{if} & a(x) > n \\ a(x) & & |a(x)| \leqq n \\ -n & & a(x) < -n \end{array}\right\} \qquad (n \geqq 1).$$

Since $a_n \in L^\infty_{real}$, we have

$$\|E[a_n]\|_{2,2} \leqq \text{Const}(1 + \|a_n\|_{BMO}) \leqq \text{Const}(1 + \|a\|_{BMO}).$$

Letting n tend to infinity, we obtain (2.38).

§2.9. Proof of (2.39)

Put

$$\hat{\sigma}_C(\alpha,\beta) = \sup\ \{\hat{\sigma}(C[a]);\ a \in L^\infty_{real},\ \alpha \leqq a \leqq \beta\} \qquad (\alpha \leqq 0,\ \ \beta \geqq 0).$$

We show that, for $\beta \geqq 1$, $0 < \delta \leqq 1$,

$$\text{(2.45)} \qquad \hat{\sigma}_C(-\beta,\beta) \leqq 2\ \hat{\sigma}_C(-\frac{\beta}{2}, \frac{\beta}{2}) + C_\delta\ \ \beta^\delta\{\sigma_C(\beta) + \beta^\delta\}\ ,$$

where

$$\sigma_C(\beta) = \sup\ \{\sigma(C[a]);\ a \in L^\infty_{real,\beta}\}\ .$$

(Tchamitchian [51] showed that

$$\sigma_C(-\beta,\beta) \leqq 2\ \sigma_C(-\frac{\beta}{2}, \frac{\beta}{2}) + \ C_\delta\ \beta^\delta \qquad (\beta \geqq 1).$$

Inequality (2.45) is an improvement of his inequality.) Here is a lemma necessary for the proof of (2.45).

Lemma 2.13. $\hat{\sigma}_C(0,\beta)=\hat{\sigma}_C(-\beta,0) \leq C_\delta\ \beta^\delta\ \{\sigma_C(\beta)+\beta^\delta\}$ $\quad(\beta \geq 1,\ 0<\delta \leq 1)$.

Proof. The first equality is evident. We define $\eta > 0$ by $\delta = \log 2/\log(1/4\eta)$. For $a \in L^\infty_{real}$, $0 \leq a \leq \beta$, $f \in L^\infty_{real,1}$, $0 \leq f \leq 1$ and an interval I, we estimate $\hat{\sigma}(I,C[a],f)$. If $(a)_I < 2\eta\beta$, we use RSL of Type 3 ($4\eta\beta - r.$, $0 - d.$). There exists $b \in L^\infty_{real}$ such that, with $\Omega = \{x \in I;\ A(x) \neq B(x)\}$,

$$0 \leq b(x) \leq 4\eta\beta \quad \text{a.e. on } I, \qquad |\Omega| \leq \frac{0+2\eta\beta}{0+4\eta\beta}\,|I| \leq |I|/2.$$

Lemma 2.8 shows that

$$\hat{\sigma}(I,C[a],f) \leq \hat{\sigma}(I,C[b],f) + \sum_{k=1}^{\infty} \hat{\sigma}(I_{\Omega,k},\ C[a],\ f)$$
$$+\ C_\delta\ \beta^\delta\ \{\sigma_C(\beta)+\beta^\delta\}|I|$$
$$\leq \{\hat{\sigma}_C(0,\ 4\eta\beta) + \frac{1}{2}\ \hat{\sigma}_C(0,\beta) + C_\delta\beta^\delta(\sigma_C(\beta)+\beta^\delta)\}\ |I|\ .$$

If $(a)_I \geq 2\eta\beta$, we use RSL of Type 1 ($\eta\beta - r.$, $\beta - a.$). There exists $b \in L^\infty_{real}$ such that, with $\Omega = \{x \in I;\ A(x) \neq B(x)\}$,

$$\eta\beta \leq b(x) \leq \beta \quad \text{a.e. on } I,\ |\Omega| \leq \frac{\beta - 2\eta\beta}{\beta - \eta\beta}\,|I| = \frac{1-2\eta}{1-\eta}\,|I|.$$

Lemma 2.8 shows that

$$\hat{\sigma}(I,C[a],f) \leq \hat{\sigma}(I,C[b],f) + \sum_{k=1}^{\infty} \hat{\sigma}(I_{\Omega,k},\ C[a],f)$$
$$+\ C_\delta\ \beta^\delta\{\sigma_C(\beta)+\beta^\delta\}\ |I|$$
$$\leq \hat{\sigma}(I,C[b],f) + \frac{1-2\eta}{1-\eta}\ \hat{\sigma}(0,\beta)\ |I|\ + C_\delta\beta^\delta\{\sigma_C(\beta)+\beta^\delta\}\ |I|.$$

We have

$$\hat{\sigma}(I,C[b],f) \leq \int_I \left|\int_I \frac{1}{(x-y)+\ i(B(x)-B(y))}\ f(y)dy\right|^2 dx$$

$$= \int_{B(I)} \left|\int_{B(I)} \frac{1}{(B^{-1}(s)-B^{-1}(t))\ +\ i(s-t)}\ \frac{f\circ B^{-1}(t)}{B'\circ B^{-1}(t)}\ dt\right|^2 \frac{ds}{B'\circ B^{-1}(s)}$$

$$\leq (\text{Const}/\eta\beta)\ \tilde{\sigma}(B(I),C[(B^{-1})'],\ f\circ B^{-1}/B'\circ B^{-1})$$

$$\leq (\text{Const}/\eta\beta)\ \text{Const}\ \{1+\|(B^{-1})'\ \chi_{B(I)}\|_\infty\ \}^{N_0}\ \|(f\circ B^{-1}/B'\circ B^{-1})\chi_{B(I)}\|^2_\infty|B(I)|$$

$$\leq C_\delta\ \beta^{-3}|B(I)| \leq C_\delta\ \beta^{-2}\ |I| \leq C_\delta|I|\ ,$$

where $B(I) = \{B(x);\ x \in I\}$ and N_0 is the absolute constant in Lemma 2.12. Hence, in this case,

$$\hat{\sigma}(I,\ C[a],f) \leq \frac{1-2\eta}{1-\eta}\ \hat{\sigma}_C(0,\beta)|I| + C_\delta\ \beta^\delta\{\sigma_C(\beta) + \beta^\delta\}|I|\ .$$

Thus we have, in both cases,

$$\frac{1}{|I|}\ \hat{\sigma}(I,\ C[a],\ f)$$
$$\leq \max\{\hat{\sigma}_C(0,4\eta\beta) + \frac{1}{2}\hat{\sigma}_C(0,\beta),\ \frac{1-2\eta}{1-\eta}\hat{\sigma}_C(0,\beta)\} + C_\delta\beta^\delta\{\sigma_C(\beta) + \beta^\delta\}\ ,$$

which yields that

$$\hat{\sigma}_C(0,\beta) \leq \max\{\hat{\sigma}_C(0,4\eta\beta) + \frac{1}{2}\hat{\sigma}_C(0,\beta),\ \frac{1-2\eta}{1-\eta}\ \hat{\sigma}_C(0,\beta)\}$$
$$+\ C_\delta\beta^\delta\{\sigma_C(\beta) + \beta^\delta\}.$$

If the second quantity in $\max\{\cdot,\cdot\}$ is larger than the first quantity, then

$$\hat{\sigma}_C(0,\beta) \leq \frac{1-\eta}{\eta}\ C_\delta\ \beta^\delta\{\sigma_C(\beta) + \beta^\delta\} = C_\delta\ \beta^\delta\{\sigma_C(\beta) + \beta^\delta\}\ .$$

If the first quantity is larger than the second quantity, then

$$\hat{\sigma}_C(0,\beta) \leq 2\ \hat{\sigma}_C(0,4\eta\beta) + C_\delta\beta^\delta\{\sigma_C(\beta) + \beta^\delta\}\ . \tag{2.46}$$

Thus, in both cases, (2.46) holds. Let m be the smallest integer of k such that $(4\eta)^k\ \beta \leq 1$. Then $m \leq \{\log\beta/\log(1/4\eta)\} + 1$. Inequality (2.46) yields that

$$\hat{\sigma}_C(0,\beta) \leq 2^m\ \hat{\sigma}_C(0,(4\eta)^m\beta) + C_\delta \sum_{k=0}^{m-1} 2^k(4\eta)^{k\delta}\beta^\delta\{\sigma_C((4\eta)^k\beta) + (4\eta)^{k\delta}\ \beta^\delta\}$$

$$\leq 2^m\ \hat{\sigma}_C(0,1) + C_\delta\ \{1 + 2^m(4\eta)^{m\delta}\}\ \beta^\delta\ \{\sigma_C(\beta) +\ \beta^\delta\}$$

$$\leq C_\delta(\beta^\delta + 2^m)\ \{\sigma_C(\beta) + \beta^\delta\} \leq C_\delta\ \beta^\delta\ \{\sigma_C(\beta) + \beta^\delta\ \}\ .$$ Q.E.D.

We now prove (2.45). For $a \in L^\infty_{real,\beta}$, $f \in L^\infty_{real}$, $0 \leq f \leq 1$ and an interval I, we estimate $\hat{\sigma}(I,C[a],f)$. Since $\hat{\sigma}(I,C[a],f) = \hat{\sigma}(I,C[-a],f)$, we may assume that $(a)_I \geq 0$. RSL of Type 1 $(-\theta\beta - \mathcal{r}.,\beta - a.)$ $(\theta = 1/2)$ shows that there exists $b \in L^\infty_{real}$ such that, with $\Omega = \{x \in I;\ A(x) \neq B(x)\}$ $= \cup_{k=1}^\infty I_k$ $(I_k = I_{\Omega,k})$,

$$- \theta\beta \leq b(x) \leq \beta \quad \text{a.e. on } I, \quad b(x) = -\theta\beta \quad (x \in \Omega),$$

$$(a)_{I_k} \leq -\theta\beta \quad (k \geq 1), \quad |\Omega| \leq \frac{\beta - (b)_I}{\beta + \theta\beta} |I| \leq \frac{|I|}{1+\theta} \qquad (\theta = 1/2).$$

Lemma 2.8 shows that

$$\hat{\sigma}(I,C[a],f) \leq \hat{\sigma}(I,C[b],f) + \sum_{k=1}^{\infty} \hat{\sigma}(I_k,C[a],f) + C_\delta \beta^\delta \{\sigma_C(\beta) + \beta^\delta\} |I|$$

$$\leq \hat{\sigma}_C(-\theta\beta,\beta)|I| + \sum_{k=1}^{\infty} \hat{\sigma}(I_k,C[a],f) + C_\delta \beta^\delta \{\sigma_C(\beta) + \beta^\delta\}|I| .$$

For each $k \geq 1$, we use RSL of Type 3 $(0 - r., -\beta - d.)$. Since $(a)_{I_k} \leq -\theta\beta$, there exist $b_k \in L^\infty_{real}$ and an open set $\Omega_k = \bigcup_{\ell=1}^{\infty} I_{k,\ell}$ $(I_{k,\ell} = I_{\Omega_{k,\ell}})$ in I_k such that

$$- \beta \leq b_k(x) \leq 0 \quad \text{a.e. on } I_k, \quad b_k(x) = 0 \quad (x \in \Omega_k),$$

$$(a)_{I_{k,\ell}} \geq 0 \quad (\ell \geq 1), \quad |\Omega_k| \leq \frac{\beta + (b_k)_{I_k}}{\beta + 0} |I_k| \leq (1-\theta)\, |I_k| .$$

Lemmas 2.8 and 2.13 show that

$$\hat{\sigma}(I_k,C[a],f) \leq \hat{\sigma}(I_k,C[b_k],f) + \sum_{\ell=1}^{\infty} \hat{\sigma}(I_{k,\ell},C[a],f) + C_\delta \beta^\delta \{\sigma_C(\beta) + \beta^\delta\} |I_k|$$

$$\leq \hat{\sigma}_C(-\beta,0)|I_k| + \sum_{\ell=1}^{\infty} \hat{\sigma}(I_{k,\ell},C[a],f) + C_\delta \beta^\delta \{\sigma_C(\beta) + \beta^\delta\} |I_k|$$

$$\leq \sum_{\ell=1}^{\infty} \hat{\sigma}(I_{k,\ell},C[a],f) + C_\delta\, \beta^\delta\, \{\sigma_C(\beta) + \beta^\delta\} |I_k|.$$

Thus

$$\hat{\sigma}(I,C[a],f) \leq \hat{\sigma}_C(-\theta\beta,\beta)|I| + \sum_{k=1}^{\infty} \sum_{\ell=1}^{\infty} \hat{\sigma}(I_{k,\ell},C[a],f)$$

$$+ \quad C_\delta \beta^\delta \{\sigma_C(\beta) + \beta^\delta\}|I|,$$

$$(a)_{I_{k,\ell}} \geq 0 \quad (k,\ell \geq 1), \quad \sum_{k=1}^{\infty} \sum_{\ell=1}^{\infty} |I_{k,\ell}| \leq \frac{1-\theta}{1+\theta} |I|.$$

Since $(a)_{I_{k,\ell}} \geq 0$, we can apply the above argument to $I_{k,\ell}$. Repeating this, we have

$$\frac{1}{|I|} \hat{\sigma}(I,C[a],f) \leq \{\hat{\sigma}_C(-\theta\beta,\beta) + C_\delta \beta^\delta (\sigma_C(\beta) + \beta^\delta)\}\{1 + \frac{1-\theta}{1+\theta} + (\frac{1-\theta}{1+\theta})^2 + \ldots\}$$

$$\leq \frac{1+\theta}{2\theta} \hat{\sigma}_C(-\theta\beta,\beta) + C_\delta \beta^\delta \{\sigma_C(\beta) + \beta^\delta\},$$

which gives

$$(2.47)\qquad \hat{\sigma}_C(-\beta,\beta) \leq \frac{1+\theta}{2\theta}\,\hat{\sigma}_C(-\theta\beta,\beta) + C_\delta\beta^\delta\,\{\sigma_C(\beta) + \beta^\delta\} \qquad (\theta = 1/2).$$

To estimate $\hat{\sigma}_C(-\theta\beta,\beta)$, we study $\hat{\sigma}(I,C[a],f)$ for $a \in L^\infty_{real}$, $-\theta\beta \leq a \leq \beta$, $f \in L^\infty_{real}$, $0 \leq f \leq 1$ and an interval I. First we assume that $(a)_I \leq 0$. RSL of Type 3 ($\theta\beta - r.$, $-\theta\beta - a.$) shows that there exists $b \in L^\infty_{real}$ such that, with $\Omega = \{x \in I;\ A(x) \neq B(x)\} = \cup_{k=1}^\infty I_k$ ($I_k = I_{\Omega,k}$),

$$-\theta\beta \leq b(x) \leq \theta\beta \quad \text{a.e. on } I, \quad b(x) = \theta\beta \quad (x \in \Omega),$$

$$(a)_{I_k} \geq \theta\beta \quad (k \geq 1), \quad |\Omega| \leq \frac{\theta\beta + (b)_I}{\theta\beta + \theta\beta}|I| \leq \frac{\theta\beta}{\theta\beta + \theta\beta}|I| = \frac{1}{2}|I|.$$

Lemma 2.8 shows that

$$\hat{\sigma}(I,C[a],f) \leq \hat{\sigma}(I,C[b],f) + \sum_{k=1}^\infty \hat{\sigma}(I_k,C[a],f) + C_\delta\beta^\delta\{\sigma_C(\beta) + \beta^\delta\}\,|I|$$

$$\leq \hat{\sigma}_C(-\theta\beta,\theta\beta)|I| + \sum_{k=1}^\infty \hat{\sigma}(I_k,C[a],f) + C_\delta\beta^\delta\{\sigma_C(\beta) + \beta^\delta\}|I|.$$

For each $k \geq 1$, we use RSL of Type 1 ($0 - r.$, $\beta - a.$). Since $(a_k)_{I_k} \geq \theta\beta$, there exist $b_k \in L^\infty_{real}$ and an open set $\Omega_k = \cup_{\ell=1}^\infty I_{k,\ell}$ ($I_{k,\ell} = I_{\Omega_k,\ell}$)) such that

$$0 \leq b_k(x) \leq \beta \quad \text{a.e. on } I_k, \quad b_k(x) = 0 \quad (x \in \Omega_k),\ (a)_{I_{k,\ell}} \leq 0 \quad (\ell \geq 1),$$

$$|\Omega_k| \leq \frac{\beta - (b_k)_{I_k}}{\beta - 0}\,|I_k| \leq \frac{\beta - \theta\beta}{\beta}\,|I_k| = (1-\theta)\,|I_k|.$$

Lemmas 2.8 and 2.13 show that

$$\hat{\sigma}(I_k,C[a],f) \leq \hat{\sigma}(I_k,C[b_k],f) + \sum_{\ell=1}^\infty \hat{\sigma}(I_{k,\ell}, C[a],f)$$

$$+ \ C_\delta\beta^\delta\,\{\sigma_C(\beta) + \beta^\delta\}|I_k| \leq \sum_{\ell=1}^\infty \hat{\sigma}(I_{k,\ell},C[a],f) + C_\delta\beta^\delta\{\sigma_C(\beta) + \beta^\delta\}|I_k|\ .$$

Thus

$$\hat{\sigma}(I,C[a],f) \leq \hat{\sigma}_C(-\theta\beta,\theta\beta)\ |I| + \sum_{k=1}^{\infty} \sum_{\ell=1}^{\infty} \hat{\sigma}(I_{k,\ell}, C[a], f)$$

$$+ \ C_\delta \beta^\delta \{\sigma_C(\beta) + \beta^\delta\}\ |I|,$$

$$(a)_{I_{k,\ell}} \leq 0 \quad (k,\ell \geq 1), \qquad \sum_{k=1}^{\infty} \sum_{\ell=1}^{\infty} |I_{k,\ell}| \leq \frac{1-\theta}{2}\ |I|.$$

Since $(a)_{I_{k,\ell}} \leq 0$, we can apply the above argument to $I_{k,\ell}$. Repeating this, we have

$$(2.48) \quad \frac{1}{|I|}\ \hat{\sigma}(I,C[a],f) \leq \{\hat{\sigma}_C(-\theta\beta,\theta\beta) + C_\delta \beta^\delta (\sigma_C(\beta) + \beta^\delta)\}\{1 + \frac{1-\theta}{2} + (\frac{1-\theta}{2})^2 + \dots\}$$

$$\leq \frac{2}{1+\theta}\ \hat{\sigma}_C(-\theta\beta,\theta\beta) + C_\delta \beta^\delta \{\sigma_C(\beta) + \beta^\delta\}\ .$$

Next we assume that $(a)_I > 0$. RSL of Type 4 ($0-r.$, $\beta-a.$) shows that there exists $b \in L^\infty_{real}$ such that, with $\Omega = \{x \in I;\ A(x) \neq B(x)\} = \cup_{k=1}^{\infty} I_k$ ($I_k = I_{\Omega,k}$),

$$0 \leq b(x) \leq \beta \quad \text{a.e. on } I, \quad (a)_{I_k} \leq 0 \quad (k \geq 0).$$

Lemmas 2.8 and 2.12 show that

$$\hat{\sigma}(I,C[a],f) \leq \hat{\sigma}(I,C[b],f) + \sum_{k=1}^{\infty} \hat{\sigma}(I_k,C[a],f) + C_\delta \beta^\delta \{\sigma_C(\beta) + \beta^\delta\}\ |I|$$

$$\leq \sum_{k=1}^{\infty} \hat{\sigma}(I_k,C[a],f) + C_\delta \beta^\delta\ \{\sigma_C(\beta) + \beta^\delta\}|I|.$$

Since $(a)_{I_k} \leq 0$, we can apply the argument in the case of $(a)_I \leq 0$ to the estimate of $\hat{\sigma}(I_k,C[a],f)$; we have

$$\hat{\sigma}(I_k,C[a],f) \leq \frac{2}{1+\theta}\ \hat{\sigma}_C(-\theta\beta,\theta\beta)|I_k| + C_\delta \beta^\delta\ \{\sigma_C(\beta) + \beta^\delta\}\ |I_k|.$$

Thus, in this case also, (2.48) holds. Consequently,

$$\hat{\sigma}_C(-\theta\beta,\beta) \leq \frac{2}{1+\theta}\ \hat{\sigma}_C(-\theta\beta,\theta\beta) + C_\delta \beta^\delta \{\sigma_C(\beta) + \beta^\delta\} \qquad (\theta = 1/2).$$

This estimate and (2.47) immediately yield (2.45).

In the same manner as in (Second Step) of the proof of (2.38), (2.45) shows that $\sigma(C[a]) \leq \mathrm{Const}(1 + \sqrt{\|a\|_\infty})$ ($a \in L^\infty_{real}$). From this inequality, we now deduce

$$(2.49) \qquad \sigma(C[a]) \leq \mathrm{Const}(1 + \sqrt{\|a\|_{BMO}}) \quad (a \text{ is real-valued}).$$

For $a \in L^\infty_{real}$, $f \in L^\infty_{real,1}$ and an interval I, we study $\sigma(I,C[a],f)$.
Let $\beta = \|a\|_{BMO}$. Inequality (2.44) holds with $E[\cdot]$ replaced by $C[\cdot]$.
Lemma 2.2 shows that there exists an open set $\Omega = \cup_{k=1}^\infty I_k$ $(I_k = I_{\Omega,k})$ in I such that

$$|\Omega| \leq \frac{1}{2\beta} \int_I |a(s) - (a)_I|\, ds \leq |I|/2, \quad (a - (a)_I)_{I_k} \leq 2\beta \qquad (k \geq 1),$$

$$|a(x) - (a)_I| \leq 2\beta \qquad \text{a.e. on } I - \Omega.$$

We put

$$b(x) = \begin{cases} a(x) & (x \in I - \Omega) \\ (a)_{I_k} & (x \in I_k,\ k \geq 1) \\ (a)_I & (x \notin \Omega). \end{cases}$$

Then

$$\sigma(I,C[a],f) \leq \sigma(I,C[b],f) + \sum_{k=1}^\infty \sigma(I_k,C[a],f) + C_\delta(1 + \beta^{2\delta})\, |I|.$$

Recall that $\|T_n[\cdot]\|_{2,2} \leq C_0^n \|\cdot\|_\infty^n$ $(n \geq 1)$ for some absolute constant C_0. (See Lemma 2.10). If $|(a)_I| \leq 4\, C_0\, \beta$, then $\|b\|_\infty \leq (2 + 4C_0)\beta$, and hence

$$\sigma(I,C[a],f) \leq \sigma_C(1 + (2 + 4C_0)\beta)\, |I| \leq \text{Const } (1 + \sqrt{\beta})\, |I|.$$

If $|(a)_I| > 4C_0\beta$, we put $\tilde{b} = b - (a)_I$. Then $\|\tilde{b}\|_\infty \leq 2\,\beta$. We have

$$\sigma(I,C[b],f) = \int_I \left| \int_I \frac{f(y)}{(x-y) - i(a)_I(x-y) + i \int_y^x \tilde{b}(s)ds}\, dy \right| dx$$

$$= \frac{1}{|1-i(a)_I|} \int_I \left|(-\pi)H(\chi_I f)(x) + \sum_{n=1}^\infty \left(\frac{-i}{1-i(a)_I}\right)^n T_n[\tilde{b}](\chi_I f)(x)\right| dx$$

$$\leq \{\pi + \sum_{n=1}^\infty (1 + |(a)_I|^2)^{-n/2}\, C_0^n\, \|\tilde{b}\|_\infty^n\}\, |I| \leq \text{Const } |I|.$$

Thus

$$\sigma(I,C[a],f) \leq \text{Const}(1 + \sqrt{\beta})\, |I| + \sum_{k=1}^\infty \sigma(I_k,C[a],f).$$

We can apply this argument to I_k. Repeating this, we have

$$\frac{1}{|I|}\sigma(I,\ C[a],\ f) \leq \text{Const } (1 + \sqrt{\beta}),$$

which gives (2.49). Lemma 2.5 and (2.49) yield (2.39). This completes the proof of (2.39).

§2.10. Application of (2.38)

As is well-known, Theorems B and C are applicable to the higher dimensional Neumann problem, pseudo-differential operators and the estimate of analytic capacity ([6], [10]). (See Chapter III.) In this section, we show an immediate application of (2.38). For a locally rectifiable curve Γ in the complex plane $\mathbb{C}$, $L^p(\Gamma)$ denotes the Banach space of functions f on Γ with norm

$$\|f\|_{L^p(\Gamma)} = \{ \int_\Gamma |f(z)|^p |dz| \}^{1/p} \qquad (1 \leq p < \infty).$$

The Cauchy(-Hilbert) transform on Γ is defined by

$$\text{(2.50)} \qquad H_\Gamma f(z) = \frac{1}{\pi} \lim_{\varepsilon \to 0} \int_{|\zeta - z| > \varepsilon} \frac{f(\zeta)}{\zeta - z} |d\zeta| .$$

The norm of H_Γ as an operator from $L^p(\Gamma)$ to itself is denoted by $\|H_\Gamma\|_{L^p(\Gamma),L^p(\Gamma)}$. We say that Γ is a chord-arc curve with constant M if, for any $z, \zeta \in \Gamma$, $\ell(z,\zeta) \leq M|z-\zeta|$, where $\ell(z,\zeta)$ is the length of Γ between z and ζ. We show

Corollary 2.14. Let Γ be a chord-arc curve with constant M. Then

$$\|H_\Gamma\|_{L^2(\Gamma),L^2(\Gamma)} \leq \text{Const } M^2.$$

Proof. Fixing $z_0 \in \Gamma$, we parametrize Γ so that

$$\Gamma = \{z(t);\ t \in \mathbb{R}\}, \quad \ell(z_0,z(t)) = |t|.$$

Let ρ be an even function in C_0^∞ such that

$$\rho(x) = 1 \quad (1 \leq x \leq M), \quad \sum_{k=0}^{3} \|\rho^{(k)}\|_\infty \leq \text{Const},$$

$$\text{supp}(\rho) \subset [-M-1, -1/2] \cup [1/2, M+1].$$

Put $h(z) = \rho(|z|)/z$ $(z \in \mathbb{C})$. Since $|s-t|/M \leq |z(s)-z(t)| \leq |s-t|$ $(s,t \in \mathbb{R})$, we have, with $a(t) = Mz'(t)$

$$\frac{1}{z(s)-z(t)} = \frac{M}{s-t} \left(\frac{Mz(s)-Mz(t)}{s-t}\right)^{-1}$$

$$= \frac{M}{s-t}\, h\left(\frac{1}{s-t} \int_t^s a(u)du\right) \ (= M\, T_{\mathbb{C}}[a,h](s,t), \text{ say}).$$

We have $|z'(t)| = 1$ a.e. and

$$\mathcal{H}_\Gamma f(z(s)) = (-\pi) \lim_{\varepsilon \to 0} \int_{|s-t|>\varepsilon} \frac{1}{z(s)-z(t)} f(z(t))z'(t)dt$$

$$= (-\pi)\, M\, T_{\mathbb{C}}[a,h]\{(f\circ z)z'\}(s) \qquad \text{a.e.}$$

Hence $\|\mathcal{H}_\Gamma\|_{L^2(\Gamma),L^2(\Gamma)} = \pi\, M\, \|T_{\mathbb{C}}[a,h]\|_{2,2}$. Let

$$\mathcal{F} h(s + it) = \int_{-\infty}^{\infty} \int_{-\infty}^{\infty} e^{-i(sx+ty)} h(x + iy)dx\, dy .$$

Then

$$T_{\mathbb{C}}[a,h] = \text{Const} \int_{-\infty}^{\infty} \int_{-\infty}^{\infty} \mathcal{F}h(s+it)\; E[\text{Re}\, \{a(s-it)\}]\; ds\, dt$$

$$= \text{Const} \int_{-\infty}^{\infty} \int_{-\infty}^{\infty} \mathcal{F}h(s+it)\; \{E[\text{Re}\, \{a(s-it)\}] + \pi H\}\; ds\, dt,$$

since $\int_{-\infty}^{\infty} \int_{-\infty}^{\infty} \mathcal{F}h(s+it)ds\, dt = \text{Const}\; h(0) = 0$. Lemma 2.10 and (2.38) show that

$$\|\, E[\text{Re}\, \{a(s-it)\}] + \pi H\|_{2,2} \leq \text{Const}\; \|\text{Re}\; a(s-it)\|_{BMO}$$

$$\leq \text{Const}\;\; M\; |s+it| ,$$

and hence

$$\|T_{\mathbb{C}}[a,h]\|_{2,2} \leq \text{Const}\; M \int_{-\infty}^{\infty} \int_{-\infty}^{\infty} |\mathcal{F}h(s+it)|\;\; |s+it|\; ds\, dt.$$

Integration by parts shows that, for $n = 2, 4$,

$$|\mathcal{F}h(s+it)| = |(-is)^{-n} \int_{-\infty}^{\infty} \int_{-\infty}^{\infty} e^{-i(sx+ty)} \frac{\partial^n}{\partial x^n} h(x+iy)\; dx\, dy|$$

$$\leq \text{Const}/|s|^n .$$

In the same manner, $|\mathcal{F}h(s + it)| \leq \text{Const}/|t|^n$ $(n = 2, 4)$. Thus $|\mathcal{F}h(s + it)| \leq \text{Const}/|s + it|^n$ $(n = 2,4)$. We have

$$\int_{-\infty}^{\infty} \int_{-\infty}^{\infty} |\mathcal{F}h(s+it)|\;\; |s+it|\; ds\, dt = \iint_{|s+it| \leq 1} + \iint_{|s+it| > 1}$$

$$\leq \text{Const}\; \{\iint_{|s+it| \leq 1} \frac{ds\, dt}{|s+it|} + \iint_{|s+it|>1} \frac{ds\, dt}{|s+it|^3}\} \leq \text{Const}.$$

Consequently,

$$\|\mathcal{H}_\Gamma\|_{L^2(\Gamma),L^2(\Gamma)} = \pi\ M\ \|T_{\mathbb{C}}[a,h]\|_2$$

$$\leq \text{Const}\ M^2 \int_{-\infty}^{\infty}\int_{-\infty}^{\infty} |\mathcal{F}h(s+it)|\ |s+it|\ ds\ dt \ \leq \text{Const}\ M^2.$$

This completes the proof of Corollary 2.24. Q.E.D.

CHAPTER III. ANALYTIC CAPACITIES OF CRANKS

§3.1. Relation between $\gamma(\cdot)$ and H

In this chapter, we study analytic capacity $\gamma(\cdot)$ from the point of view of integralgeometry and the Cauchy transform on graphs. We shall estimate $\gamma(\cdot)$ of so-called cranks. For a compact set E in the complex plane $\mathbb{C}$, $H^\infty(E^c)$ denotes the Banach space of bounded analytic functions in $\mathbb{C} \cup \{\infty\} - E (= E^c)$ with supremum norm $\|\cdot\|_{H^\infty}$. The analytic capacity of E is defined by

$$(3.1) \qquad \gamma(E) = \sup\{|f'(\infty)| \ ; \ \|f\|_{H^\infty} \leqq 1, \ f \in H^\infty(E^c)\},$$

where $f'(\infty) = \lim_{z \to \infty} z(f(z) - f(\infty))$, i.e., $f'(\infty)$ is the $1/z$-coefficient of the Taylor expansion at ∞ ([29, p.6]). If $f \in H^\infty(E^c)$, $\|f\|_{H^\infty} \leqq 1$, then

$$g(z) = (f(z) - f(\infty))/(1 - \overline{f(\infty)}\, f(z)) \quad (\in H^\infty(E^c))$$

satisfies $g(\infty) = 0$, $\|g\|_{H^\infty} \leqq 1$ and

$$|g'(\infty)| = |f'(\infty)|/(1 - |f(\infty)|^2) \geqq |f'(\infty)|.$$

Hence, to estimate $\gamma(\cdot)$, we can restrict our attention to functions vanishing at ∞. The Cauchy transform of a complex measure μ in $\mathbb{C}$ is defined by

$$C\mu(z) = \frac{1}{2\pi i} \int_{\mathbb{C}} \frac{1}{\zeta - z}\, d\mu(\zeta) \qquad (z \notin \mathrm{supp}(\mu)).$$

We put

$$(3.2) \quad \gamma_+(E) = \sup\{\frac{1}{2\pi} \int d\mu; \ \|C\mu\|_{H^\infty} \leqq 1, \ \mu \geqq 0, \ \mathrm{supp}(\mu) \subset E\}.$$

Since $(C\mu)'(\infty) = -\frac{1}{2\pi i} \int d\mu$, we have $\gamma(E) \geqq \gamma_+(E)$. Let $D(z,r)$ denote the open disk of center z and of radius r. For $\varepsilon > 0$, we put $|E|_\varepsilon = 2 \inf \Sigma_{k=1}^\infty r_k$, where the infimum is taken over all coverings $\{D(z_k,r_k)\}_{k=1}^\infty$ of E with radii less than ε. The generalized length of E is defined by $|E| = \lim_{\varepsilon \to 0} |E|_\varepsilon$. If $E \subset \mathbb{R}$, then the generalized length of E equals its 1-dimension Lebesgue measure. We shall compare $\gamma(\cdot)$, $\gamma_+(\cdot)$ and $|\cdot|$.

A set $\Gamma \subset \mathbb{C}$ is called a locally chord-arc curve with constant M, if, for any $z \in \Gamma$, there exists $\varepsilon > 0$ such that $\Gamma \cap D(z,\varepsilon)$ is a chord-arc curve with constant M. A locally chord-arc curve is not, in general, connected. Let Γ be a locally chord-arc compact curve with constant 100. We define

(3.3) $\rho(\Gamma) = \inf \gamma(E)/|E|, \quad \rho_+(\Gamma) = \inf \gamma_+(E)/|E|,$

where the infimums are taken over all compact sets E on Γ. Let $L^p(\Gamma)$, $\|\cdot\|_{L^p(\Gamma)}$ $(1 \leqq p < \infty)$ be the same as in §2.10. Let $L^\infty(\Gamma)$ be the L^∞ space on Γ with supremum norm $\|\cdot\|_{L^\infty(\Gamma)}$, and let $L^1_w(\Gamma)$ be the space of functions f on Γ with norm

$$\|f\|_{L^1_w(\Gamma)} = \sup \{\lambda \bigl| z \in \Gamma;\ |f(z)| > \lambda \bigr| \ ;\ \lambda > 0\}.$$

The Cauchy transform H_Γ on Γ is defined by (2.50). The norm of H_Γ as an operator from $L^1(\Gamma)$ to $L^1_w(\Gamma)$ is denoted by $\|H_\Gamma\|_{L^1(\Gamma),L^1_w(\Gamma)}$. Here are relations among $\rho(\Gamma)$, $\rho_+(\Gamma)$ and $\|H_\Gamma\|_{L^1(\Gamma),L^1(\Gamma)}$.

<u>Theorem D</u>.

(3.4) $\mathrm{Const}/\|H_\Gamma\|_{L^1(\Gamma),L^1_w(\Gamma)} \leqq \rho_+(\Gamma) \leqq \mathrm{Const}/\|H_\Gamma\|_{L^1(\Gamma),L^1_w(\Gamma)},$

(3.5) $\rho_+(\Gamma) \leqq \rho(\Gamma) \leqq \mathrm{Const}\ \rho_+(\Gamma)^{1/3}.$

We begin by showing the second inequality in (3.4). Let $f \in L^2(\Gamma)$, $\|f\|_{L^1(\Gamma)} = 1$. For $\lambda > 0$, $-\pi < \theta \leqq \pi$, we put

$$E_{\lambda,\theta} = \{z \in \Gamma;\ H_\Gamma f(z) \in D(\lambda e^{i\theta}, \lambda/4)\}.$$

There exists a compact set $F_{\lambda,\theta}$ in $E_{\lambda,\theta}$ such that $|F_{\lambda,\theta}| \geqq |E_{\lambda,\theta}|/2$. There exists a non-negative measure μ on $F_{\lambda,\theta}$ such that

$$\|C\mu\|_{H^\infty} \leqq 1, \quad \frac{1}{2\pi} \int_{F_{\lambda,\theta}} d\mu \geqq \gamma_+(F_{\lambda,\theta})/2.$$

Since $\|C\mu\|_{H^\infty} \leqq 1$, we can write $d\mu = h|dz|$ with $h \in L^\infty(\Gamma)$, $0 \leqq h \leqq 2\pi$. We show that $\|H_\Gamma h\|_{L^\infty(\Gamma)} \leqq \mathrm{Const}$.

For $z_0 \in \Gamma$ such that $H_\Gamma h(z_0)$ exists (in the sense of (2.50)), we choose first $\varepsilon > 0$ so that $\Gamma \cap D(z_0,\varepsilon)$ is a chord-arc curve with constant 100. Choose next $0 < \varepsilon' < \varepsilon$ so that, for any $z \in \Gamma^c \cap D(z_0,\varepsilon')$,

$$|H_\Gamma h(z_0) - 2i\, C\mu(z)| \leqq |H_\Gamma \tilde{h}(z_0) - 2i\, C(\tilde{h}|d\zeta|)(z)| + 1,$$

where $\tilde{h}$ is the restriction of h to $\Gamma \cap D(z_0,\varepsilon)$. Choose at last $0 < \varepsilon'' < \varepsilon'$ so that

$$\left| H_\Gamma \tilde{h}(z_0) - \frac{1}{\pi} \int_{\Gamma \cap D(z_0,\varepsilon'')^c} \frac{\tilde{h}(\zeta)}{\zeta - z_0} |d\zeta| \right| \leq 1.$$

Since $\Gamma \cap D(z_0,\varepsilon)$ is a chord-arc curve with constant 100, there exists $z_0' \in \mathbb{C}$ such that $\varepsilon''/4 \leq |z_0 - z_0'| \leq \varepsilon''/2$ and $\Gamma \cap D(z_0', 10^{-10}\varepsilon'') = \emptyset$. Thus we have

$$|H_\Gamma h(z_0) - 2i\, C\, \mu(z_0')| \leq |H_\Gamma \tilde{h}(z_0) - 2i\, C(\tilde{h}|d\zeta|)(z_0')| + 1$$

$$\leq \left| \frac{1}{\pi} \int_{\Gamma \cap D(z_0,\varepsilon'')^c} \frac{\tilde{h}(\zeta)}{\zeta - z_0} |d\zeta| - \frac{1}{\pi} \int_\Gamma \frac{\tilde{h}(\zeta)}{\zeta - z_0'} |d\zeta| \right| + 2$$

$$\leq 2 \int_{\Gamma \cap (D(z_0,\varepsilon) - D(z_0,\varepsilon''))} \frac{|z_0 - z_0'|}{|\zeta - z_0||\zeta - z_0'|} |d\zeta|$$

$$+ 2 \int_{\Gamma \cap D(z_0,\varepsilon'')} \frac{1}{|\zeta - z_0'|} |d\zeta| + 2 \leq \text{Const.}$$

Since $\|C\mu\|_{H^\infty} \leq 1$, we have $|H_\Gamma h(z_0)| \leq$ Const. Since $H_\Gamma h(z)$ exists a.e. on Γ, $\|H_\Gamma h\|_{L^\infty(\Gamma)} \leq$ Const.

Since

$$\frac{1}{2} \gamma_+(F_{\lambda,\theta}) \leq \frac{1}{2\pi} \int_{F_{\lambda,\theta}} h|dz| \leq \gamma_+(F_{\lambda,\theta}),$$

we have

$$\frac{\lambda}{2} \gamma_+(F_{\lambda,\theta}) \leq \frac{1}{2\pi} \left| \int_{F_{\lambda,\theta}} \lambda e^{i\theta} h|dz| \right|$$

$$\leq \frac{1}{2\pi} \left| \int_{F_{\lambda,\theta}} (H_\Gamma f)\, h\, |dz| \right| + \frac{1}{8\pi} \int_{F_{\lambda,\theta}} h|dz|$$

$$\leq \frac{1}{2\pi} \left| \int_\Gamma f(H_\Gamma h)|dz| \right| + \frac{\lambda}{4} \gamma_+(F_{\lambda,\theta})$$

$$\leq \text{Const} \|f\|_{L^1(\Gamma)} + \frac{\lambda}{4} \gamma_+(F_{\lambda,\theta}),$$

which gives

$$\gamma_+(F_{\lambda,\theta}) \leq \text{Const} \|f\|_{L^1(\Gamma)}/\lambda \leq \text{Const}/\lambda .$$

Since

$$|E_{\lambda,\theta}| \leq 2|F_{\lambda,\theta}| \leq 2\, \gamma_+(F_{\lambda,\theta})/\rho_+(\Gamma) \leq \text{Const}/(\lambda\, \rho_+(\Gamma)),$$

we have

$$|z \in \Gamma;\ |H_\Gamma f(z)| > \lambda| \leq \left| \bigcup_{n=0}^{\infty} \bigcup_{k=1}^{100} E_{(5/4)^n\lambda,\ 2\pi k/100} \right|$$

$$\leq \{\mathrm{Const}/(\lambda\ \rho_+(\Gamma))\} \sum_{n=0}^{\infty} (\tfrac{4}{5})^n = \mathrm{Const}/(\lambda\ \rho_+(\Gamma)).$$

Since $\lambda > 0$ is arbitrary,

$$\|H_\Gamma f\|_{L^1_w(\Gamma)} \leq \mathrm{Const}/\ \rho_+(\Gamma) \qquad (f \in L^2(\Gamma),\ \|f\|_{L^1(\Gamma)} = 1).$$

A standard argument shows that this inequality holds with f replaced by any $g \in L^1(\Gamma)$, $\|g\|_{L^1(\Gamma)} = 1$. Hence the second inequality in (3.4) holds.

The proof of the first inequality in (3.4) was essentially given by Davie [21], Marshall [37] and Davie-Øksendal [22]. Here is a tool for the proof.

Lemma 3.1 (the separation theorem [53, p. 108]). Let P, Q be two compact convex sets in the Banach space $C(\Gamma)$ of continuous functions on Γ with norm $\|\cdot\|_{L^\infty(\Gamma)}$. Then there exists a complex measure μ on Γ such that, for any $f \in P$, $g \in Q$,

$$\mathrm{Re} \int_\Gamma f\, d\mu > \mathrm{Re} \int_\Gamma g\, d\mu.$$

Since Γ is a locally chord-arc compact curve with constant 100, there exists $\varepsilon_0 > 0$ such that $\Gamma \cap D(z,\varepsilon_0)$ is a chord-arc curve with constant 100 for any $z \in \Gamma$. Let H_Γ^ε $(0 < \varepsilon < \varepsilon_0/2)$ be an operator defined by

$$H_\Gamma^\varepsilon f(z) = \frac{1}{\pi} \int_{\Gamma \cap D(z,\varepsilon)^c} \frac{f(\zeta)}{\zeta - z}\, |d\zeta| \qquad (f \in L^1(\Gamma)).$$

We show that there exists an absolute constant C_0 such that

$$\text{(3.6)} \quad \|H_\Gamma^\varepsilon\|_{L^1(\Gamma),L^1_w(\Gamma)} \leq C_0\, \|H_\Gamma\|_{L^1(\Gamma),L^1_w(\Gamma)} \qquad (= C_0 m_0,\ \text{say}).$$

Let $M_\Gamma^{2\varepsilon}$ be a maximal operator defined by

$$M_\Gamma^{2\varepsilon} f(z) = \sup_{0 < \eta \leq 2\varepsilon} \frac{1}{|\Gamma(z,\eta)|} \int_{\Gamma(z,\eta)} |f(\zeta)|\, |d\zeta| \qquad (f \in L^1(\Gamma)),$$

where $\Gamma(z,\eta) = \Gamma \cap D(z,\eta)$. Then

$$\|M_\Gamma^{2\varepsilon}\|_{L^1(\Gamma),L^1_w(\Gamma)} \leq \mathrm{Const}.$$

For $f \in L^1(\Gamma)$ with $\|f\|_{L^1(\Gamma)} \leq 1$, we put

$$E = \{z \in \Gamma;\ |H_\Gamma f(z)| \leq \lambda\ ,\ M_\Gamma^{2\varepsilon} f(z) \leq \lambda\}\ .$$

Then

$$|\Gamma - E| \leq \{\|H_\Gamma\|_{L^1(\Gamma),L^1_w(\Gamma)} + \|M_\Gamma^{2\varepsilon}\|_{L^1(\Gamma),L^1_w(\Gamma)}\} / \lambda$$

$$\leq (m_0 + \text{Const})/\lambda \leq \text{Const } m_0/\lambda.$$

Let $z_0 \in E$. Then, for any $z \in E \cap D(z_0,\varepsilon/2)$,

$$|H_\Gamma^\varepsilon f(z)| \leq |H_\Gamma^\varepsilon f(z) - H_\Gamma f(z)| + |H_\Gamma f(z)|$$

$$\leq \frac{1}{\pi} \left|\int_{\Gamma(z,\varepsilon)} \frac{f(\zeta)}{\zeta - z} |d\zeta|\right| + \lambda$$

$$\leq \frac{1}{\pi} \left|\int_{\Gamma(z_0,\varepsilon)} \frac{f(\zeta)}{\zeta - z} |d\zeta|\right| + \text{Const } M_\Gamma^{2\varepsilon} f(z) + \lambda$$

$$\leq \frac{1}{\pi} \left|\int_{\Gamma(z_0,\varepsilon)} \frac{f(\zeta)}{\zeta - z} |d\zeta|\right| + \text{Const } \lambda .$$

Hence we have

$$|H_\Gamma^\varepsilon f(z)| \leq |H_{\Gamma(z_0,\varepsilon)} f_0(z)| + C_0' \lambda \qquad (z \in E \cap \Gamma(z_0,\varepsilon/2)),$$

where f_0 is the restriction of f to $\Gamma(z_0,\varepsilon)$ and C_0' is an absolute constant. Since $\Gamma(z_0,\varepsilon)$ is a chord-arc curve with constant 100, Corollary 2.14 shows that $\|H_{\Gamma(z_0,\varepsilon)}\|_{L^1(\Gamma),L^1_w(\Gamma)} \leq \text{Const}$. (See also (2.10).) Thus

$$|z \in E \cap \Gamma(z_0,\varepsilon/2);\ |H_\Gamma^\varepsilon f(z)| > (C_0' + 1)\lambda|$$

$$\leq |z \in E \cap \Gamma(z_0,\varepsilon/2);\ |H_{\Gamma(z_0,\varepsilon)} f_0(z)| > \lambda| \leq (\text{Const}/\lambda)\|f_0\|_{L^1(\Gamma(z_0,\varepsilon))} .$$

We choose a finite covering $\{D(z_k,\varepsilon/2)\}_{k=1}^n$ of Γ so that $z_k \in \Gamma$ $(1 \leq k \leq n)$ and $\|\sum_{k=1}^n \chi_{\Gamma(z_k,\varepsilon)}\|_{L^\infty(\Gamma)} \leq \text{Const}$, where $\chi_{\Gamma(z_k,\varepsilon)}$ is the characteristic function of $\Gamma(z_k,\varepsilon)$. Then

$$|z \in \Gamma;\ |H_\Gamma^\varepsilon f(z)| > (C_0' + 1)\lambda|$$

$$\leq |z \in E; |H_\Gamma^\varepsilon f(z)| > (C_0' + 1)\lambda| + |\Gamma - E|$$

$$\leq \sum_{k=1}^n |z \in E \cap \Gamma(z_k,\varepsilon/2); |H_\Gamma^\varepsilon f(z)| > (C_0' + 1)\lambda| + \text{Const } m_0/\lambda$$

$$\leq (\text{Const}/\lambda) \sum_{k=1}^n \int_{\Gamma(z_k,\varepsilon)} |f(\zeta)|\,|d\zeta| + \text{Const } m_0/\lambda \leq \text{Const } m_0/\lambda,$$

which gives (3.6).

Given $0 < \varepsilon < \varepsilon_0/2$ and a compact set $E \subset \Gamma$, we put

$$F = \{f \in L^\infty(\Gamma);\ 0 \leq f(z) \leq 1,$$

$$\int_E f(z)|dz| \geq |E|/2,\quad \mathrm{supp}(f) \subset E\},$$

$$P = \{H_\Gamma^\varepsilon f;\ f \in F\},\qquad Q = \{g \in C(\Gamma);\ \|g\|_{L^\infty(\Gamma)} \leq 3\, C_0 m_0\},$$

where $m_0 = \|H_\Gamma\|_{L^1(\Gamma), L^1_w(\Gamma)}$ and C_0 is the constant in (3.6). We show that $P \cap Q \neq \emptyset$. Suppose that $P \cap Q = \emptyset$. Since P, Q are compact and convex in $C(\Gamma)$, Lemma 3.1 shows that there exists a measure μ on Γ such that

$$\mathrm{Re} \int_\Gamma H_\Gamma^\varepsilon f \ d\mu > \mathrm{Re} \int_\Gamma g\ d\mu \qquad (f \in F,\ g \in Q).$$

Taking the supremum of $\mathrm{Re} \int_\Gamma g\ d\mu$ over all $g \in Q$, we have

$$\mathrm{Re} \int_\Gamma H_\Gamma^\varepsilon f\ d\mu \geq 3\, C_0 m_0 \int_\Gamma |d\mu| \qquad (f \in F),$$

which implies that

$$-\mathrm{Re} \int_\Gamma f\, g_0\ |dz| \geq 3\, C_0 m_0 \qquad (f \in F),$$

where

$$g_0(z) = (\pi \int_\Gamma |d\mu|)^{-1} \int_{\Gamma, |\zeta - z| > \varepsilon} \frac{1}{\zeta - z}\ d\mu(\zeta).$$

By (3.6), we have, for any $h \in L^1(\Gamma)$ with $\|h\|_{L^1(\Gamma)} \leq 1$,

$$|z \in \Gamma; |H_\Gamma^\varepsilon h(z)| \geq 2\, C_0 m_0/|E|| \leq |E|/2.$$

Since the kernel of H_Γ^ε is uniformly bounded, this inequality holds with $h\ |dz|$ replaced by any measure μ with $\int_\Gamma |d\mu| \leq 1$. Hence

$$|z \in \Gamma;\ |g_0(z)| \geq 2\, C_0 m_0/|E|| \leq |E|/2.$$

Let $F = \{z \in E;\ |g_0(z)| \leq 2\, C_0 m_0/|E|\}$ and let χ_F be its characteristic function. Then $\chi_F \in F$. Hence we have

$$3\, C_0 m_0 \leq -\mathrm{Re} \int_\Gamma \chi_F\, g_0\ |dz| \leq \int_F |g_0(z)|\ |dz|$$

$$\leq \frac{2\, C_0 m_0}{|E|} \int_F |dz| \leq 2\, C_0 m_0,$$

which is a contradiction. Thus $P \cap Q \neq \emptyset$.

Since $P \cap Q \neq \emptyset$, there exists $f^\varepsilon \in L^\infty(\Gamma)$ such that $f^\varepsilon \in F$, $\|H_\Gamma^\varepsilon f^\varepsilon\|_{L^\infty(\Gamma)} \leq 3\, C_0 m_0$. Let $\{\varepsilon_n\}_{n=1}^\infty$ be a sequence of positive numbers such that $\varepsilon_1 \geq \varepsilon_2 \geq \dots$, $\lim_{n\to\infty} \varepsilon_n = 0$ and $\{f^{\varepsilon_n}|dz|\}_{n=1}^\infty$ converges weakly (as a sequence of measures). Then the limit is absolutely continuous with respect to $|dz|$; we write by $f^0|dz|$. We have $f^0 \in F$. We show that $\|H_\Gamma f^0\|_{L^\infty(\Gamma)} \leqq \text{Const}\, m_0$. Let $z_0 \in \Gamma$ and let $\varepsilon_k, \varepsilon_\ell$ satisfy $0 < 2\varepsilon_\ell < \varepsilon_k < \varepsilon_0/2$. Then, for any $z \in \Gamma(z_0,\varepsilon_k/2)$,

$$|H_\Gamma^{\varepsilon_k} f^{\varepsilon_\ell}(z_0)| \leqq \frac{1}{\pi} \left|\int_{\Gamma - \Gamma(z_0,\varepsilon_k)} \frac{f^{\varepsilon_\ell}(\zeta)}{\zeta - z} |d\zeta|\right|$$

$$+ \frac{1}{\pi} \left|\int_{\Gamma,\ \varepsilon_k < |\zeta - z_0| < \varepsilon_0/2} \left(\frac{1}{\zeta - z_0} - \frac{1}{\zeta - z}\right) f^{\varepsilon_\ell}(\zeta)\, |d\zeta|\right|$$

$$+ \frac{1}{\pi} \left|\int_{\Gamma - \Gamma(z_0,\varepsilon_0/2)} \left(\frac{1}{\zeta - z_0} - \frac{1}{\zeta - z}\right) f^{\varepsilon_\ell}(\zeta)\, |d\zeta|\right|$$

$$\leqq \frac{1}{\pi} \left|\int_{\Gamma - \Gamma(z_0,\varepsilon_k)} \frac{f^{\varepsilon_\ell}(\zeta)}{\zeta - z} |d\zeta|\right| + \text{Const}\ M_\Gamma^{\varepsilon_0} f^{\varepsilon_\ell}(z_0)$$

$$+ \text{Const}\ (\varepsilon_k/\varepsilon_0^2) \int_{\Gamma - \Gamma(z_0,\varepsilon_0/2)} f^{\varepsilon_\ell}(\zeta)\, |d\zeta|$$

$$\leqq \frac{1}{\pi} \left|\int_{\Gamma - \Gamma(z_0,\varepsilon_k)} \frac{f^{\varepsilon_\ell}(\zeta)}{\zeta - z} |d\zeta|\right| + \text{Const}\ \{1 + (\varepsilon_k/\varepsilon_0^2)|\Gamma|\}.$$

Let $f_k^{\varepsilon_\ell}$ denote the restriction of f^{ε_ℓ} to $\Gamma(z_0,\varepsilon_k)$. Then, for any $z \in \Gamma(z_0,\varepsilon_k/2)$,

$$\frac{1}{\pi} \left|\int_{\Gamma - \Gamma(z_0,\varepsilon_k)} \frac{f^{\varepsilon_\ell}(\zeta)}{\zeta - z} |d\zeta|\right| = |H_\Gamma f^{\varepsilon_\ell}(z) - H_\Gamma^{\varepsilon_\ell} f_k^{\varepsilon_\ell}(z)|$$

$$\leqq 3\, C_0 m_0 + |H_\Gamma^{\varepsilon_\ell} f_k^{\varepsilon_\ell}(z)|,$$

which shows that

$$|H_\Gamma^{\varepsilon_k} f^{\varepsilon_\ell}(z_0)| \leqq |H_\Gamma^{\varepsilon_\ell} f_k^{\varepsilon_\ell}(z)| + 3\, C_0 m_0$$

$$+ \text{Const}\ \{1 + (\varepsilon_k/\varepsilon_0^2)|\Gamma|\} \qquad (z \in \Gamma(z_0,\varepsilon_k/2)).$$

By (3.6), we have

$$|z \in \Gamma;\ |H_\Gamma^{\varepsilon_\ell} f_k^{\varepsilon_\ell}(z)| \geqq 10^3\, C_0 m_0| \leqq 10^{-3} \|f_k^{\varepsilon_\ell}\|_{L^1(\Gamma)}$$

$$\leqq 10^{-3} |\Gamma(z_0,\varepsilon_k)| \leqq \frac{1}{2} |\Gamma(z_0,\varepsilon_k/2)|.$$

This shows that the generalized length of $\{z \in \Gamma(z_0,\varepsilon_k/2);\ |H_\Gamma^{\varepsilon_\ell} f_k^{\varepsilon_\ell}(z)| \leqq 10^3 C_0 m_0\}$ is larger than or equal to $|\Gamma(z_0,\varepsilon_k/2)|/2$. Hence the mean of

$|H_\Gamma^{\varepsilon_\ell} f_k^{\varepsilon_\ell}(z)|$ over this set is dominated by Const m_0, which shows that

$$|H_\Gamma^{\varepsilon_k} f^{\varepsilon_\ell}(z_0)| \leq \text{Const } m_0 + 3\, C_0 m_0 + \text{Const }\{1 + (\varepsilon_k/\varepsilon_0^2)|\Gamma|\}$$

$$\leq \text{Const }\{m_0 + (\varepsilon_k/\varepsilon_0^2)|\Gamma|\}.$$

Since $z_0 \in \Gamma$ is arbitrary,

$$\|H_\Gamma^{\varepsilon_k} f^{\varepsilon_\ell}\|_{L^\infty(\Gamma)} \leq \text{Const }\{m_0 + (\varepsilon_k/\varepsilon_0^2)|\Gamma|\}.$$

Letting first ℓ tend to infinity, and letting next k tend to infinity, we have $\|H_\Gamma f^0\|_{L^\infty(\Gamma)} \leq \text{Const } m_0$.

Now let $d\mu^0 = f^0|dz|$. Since $f^0 \in F$, $\|H_\Gamma f^0\|_{L^\infty(\Gamma)} \leq \text{Const } m_0$, the maximum modulus principle shows that

$$\|C\mu^0\|_{H^\infty} \leq C_1 m_0, \quad \frac{1}{2\pi}\int_E d\mu^0 \geq |E|/C_1, \quad \text{supp}(\mu^0) \subset E,$$

where C_1 is an absolute constant. Let $\mu^{00} = \mu^0/(C_1 m_0)$. Then

$$\mu^{00} \geq 0, \ \text{supp}(\mu^{00}) \subset E, \ \|C\mu^{00}\|_{H^\infty} \leq 1.$$

Hence

$$\gamma_+(E) \geq \frac{1}{2\pi}\int_E d\mu^{00} \geq |E|/(C_1^2 m_0),$$

which shows that $\gamma_+(E)/|E| \geq \text{Const } m_0$. Taking the infimum over all compact sets $E \subset \Gamma$, we obtain the first inequality in (3.4).

The first inequality in (3.5) is evident. At last we show the second inequality in (3.5). Let $f \in L^2(\Gamma)$, $\|f\|_{L^1(\Gamma)} = 1$. For $\lambda > 0$, $-\pi < \theta \leq \pi$, we put

$$E_{\lambda,\theta} = \{z \in \Gamma;\ H_\Gamma f(z) \in D(\lambda e^{i\theta}, \rho\lambda/4)\} \qquad (\rho = \rho(\Gamma)).$$

There exists a compact set $F_{\lambda,\theta}$ in $E_{\lambda,\theta}$ such that $|F_{\lambda,\theta}| \geq |E_{\lambda,\theta}|/2$. There exists $g \in H^\infty(F^c_{\lambda,\theta})$ such that

$$\|g\|_{H^\infty} \leq 1, \quad g(\infty) = 0, \quad |g'(\infty)| \geq \gamma(F_{\lambda,\theta})/2.$$

We can write $g = C(h|dz|)$ with $h \in L^\infty(\Gamma)$ satisfying

$$\|h\|_{L^\infty(\Gamma)} \leq 2\pi, \quad \|H_\Gamma h\|_{L^\infty(\Gamma)} \leq \text{Const}, \quad \text{supp}(h) \subset F_{\lambda,\theta}.$$

Since

$$\frac{1}{2}\gamma(F_{\lambda,\theta}) \leq |g'(\infty)| = \frac{1}{2\pi}\left|\int_{F_{\lambda,\theta}} h|dz|\right| \leq \gamma(F_{\lambda,\theta}),$$

we have

$$\frac{\lambda}{2}\gamma(F_{\lambda,\theta}) \leq \frac{1}{2\pi}\left|\int_{F_{\lambda,\theta}} \lambda e^{i\theta} h|dz|\right|$$

$$\leq \frac{1}{2\pi}\left|\int_{F_{\lambda,\theta}} (H_\Gamma f)\, h\, |dz|\right| + \frac{\rho\lambda}{8\pi}\int_{F_{\lambda,\theta}} |h||dz|$$

$$\leq \frac{1}{2\pi}\left|\int_\Gamma f\, H_\Gamma\, h\, |dz|\right| + \frac{\rho\lambda}{4}|F_{\lambda,\theta}|$$

$$\leq \text{Const}\, \|f\|_{L^1(\Gamma)} + \frac{\lambda}{4}\gamma(F_{\lambda,\theta}),$$

which gives $\gamma(F_{\lambda,\theta}) \leq \text{Const}\, \|f\|_{L^1(\Gamma)}/\lambda = \text{Const}/\lambda$. Since

$$|E_{\lambda,\theta}| \leq 2|F_{\lambda,\theta}| \leq \gamma(F_{\lambda,\theta})/\rho \leq \text{Const}/(\rho\lambda),$$

we have, with $q =$ (the integral part of $10^3/\rho$),

$$|z \in \Gamma;\ |H_\Gamma f(z)| > \lambda| \leq \left|\bigcup_{n=0}^{\infty}\bigcup_{k=1}^{q} E_{\{1+(\rho/4)\}^n\lambda,\,2\pi k/q}\right|$$

$$\leq \text{Const}/(\rho^3\lambda),$$

which gives that

$$\|H_\Gamma\|_{L^1(\Gamma),L^1_w(\Gamma)} \leq \text{Const}/\rho(\Gamma)^3.$$

This inequality and the first inequality in (3.4) immediately yield the second inequality in (3.5).

§3.2. Vitushkin's example, Garnett's example, Calderón's problem and extremal problems ([5], [28], [46], [52])

Painlevé showed that the analytic capacity of a compact set of zero generalized length is equal to zero. For a compact set E of finite positive generalized length, we can choose a finite covering $\{D(z_k,r_k)\}_{k=1}^n$ of E so that $\Sigma_{k=1}^n r_k \leq |E|$. Then, for any $f \in H^\infty(E^c)$ with $\|f\|_{H^\infty} \leq 1$,

$$|f'(\infty)| = \frac{1}{2\pi}\left|\int_{\partial\{\bigcup_{k=1}^n D(z_k,r_k)\}} f(\zeta)\, d\zeta\right|$$

$$\leq \text{Const} \sum_{k=1}^{n} r_k \leq \text{Const}\, |E|,$$

which gives

(3.7) $\gamma(E) \leq \text{Const}|E|$.

This inequality immediately yields Painlevé's theorem. Vitushkin [52] constructed an example P_∞ such that $\gamma(P_\infty) = 0$, $|P_\infty| > 0$. The set P_∞ is defined as follows. Let $P_0 = [0,1]$. We divide P_0 into two non-overlapping closed segments $[0,1/2]$, $[1/2,1]$. Fixing their midpoints, we rotate these two segments so that the resulting two segments are perpendicular to the x-axis. (The midpoints of the resulting segments are on $\mathbb{R}$.) Let P_1 be the union of these two segments. We divide each segment (of P_1) into two non-overlapping closed segments of equal length. Fixing their midpoints, we rotate these four segments so that the resulting four segments are perpendicular to the y-axis. Repeating this discussion, we define P_n; P_n is a union of 2^n closed segments of length 2^{-n}. We put $P_\infty = \bigcap_{n=0}^\infty \overline{\bigcup_{k=n}^\infty P_k}$. Garnett [28] also constructed an example Q_∞ such that $\gamma(Q_\infty) = 0$, $|Q_\infty| > 0$. The set Q_∞ is defined as follows. Let $Q_0 = [0,1] \times [0,1]$. Let Q_1 be the union of four closed squares with sides of length $1/4$ in the four corners of Q_0. In the same manner, Q_n is defined from Q_{n-1} with each component of Q_{n-1} replaced by four closed squares with sides of length 4^{-n} in the four corners of the component. We put $Q_\infty = \bigcap_{n=0}^\infty Q_n$.

There are several proofs of $\gamma(Q_\infty) = 0$. Supposing that the non-trivial Ahlfors function [29, p. 18] of Q_∞ exists, Garnett [28] showed a contradiction. Using Besicovitch's set theory [1], Mattila [38] also gave an indirect proof. A direct proof from the point of view of the construction of Garabedian functions [29, p. 19] is given in [46]. This method is applicable to estimate $\gamma(\cdot)$ of various sets.

It is sufficient to construct a sequence $\{(R_n, f_n)\}_{n=10000}^\infty$ of pairs of compact sets R_n and $f_n \in H^\infty(R_n^c)$ so that $R_n \supset Q_n$, $f_n(\infty) = 1$ and

$$\int_{\partial R_n} |f_n(z)|\,|dz| \leq \text{Const}/(\log n).$$

If such a sequence exists, we have, for any $g \in H^\infty(Q_n^c)$, $\|g\|_{H^\infty} \leq 1$, $g(\infty) = 0$,

$$|g'(\infty)| = \frac{1}{2\pi}\left|\int_{\partial R_n} g(z)\,dz\right| = \frac{1}{2\pi}\left|\int_{\partial R_n} g(z) f_n(z)\,dz\right|$$

$$\leq \frac{1}{2\pi}\int_{\partial R_n} |f_n(z)||dz| \leq \text{Const}/(\log n),$$

which shows that $\gamma(Q_n) \leq \text{Const}/(\log n)$. Thus $\gamma(Q_\infty) = 0$. The pair (R_n, f_n) is constructed as follows. We denote by m the integral part of $(\log n)/2$. For a closed square Q with sides parallel to the coordinate axes, $\zeta(Q)$ denotes its lower left corner and $\ell(Q)$ denotes the length of a side. Let Q_-, Q_+ denote two closed squares in Q with sides parallel to the coordinate axes such that $\ell(Q_-) = \ell(Q_+) = 4^{-m}\ell(Q)$, $\zeta(Q_-) = \zeta(Q)$ and the upper right corners of Q_+ and Q are identical. For $k \geqq 0$, we can write $Q_k = \bigcup_{j=1}^{4^k} Q_{k,j}$ with components $\{Q_{k,j}\}_{j=1}^{4^k}$. Note that $\ell(Q_{k,j}) = 4^{-k}$. We define inductively $4^m + 1$ compact sets $\{V_k\}_{k=0}^{4^m}$ by $V_0 = (Q_0)_- \cup (Q_0)_+$,

$$V_k = \bigcup_{\ell \in \Omega_k} (Q_{km,\ell})_- \cup (Q_{km,\ell})_+ \qquad (1 \leqq k \leqq 4^m),$$

where $\Omega_k = \{1 \leqq \ell \leqq 4^{km};\ Q_{km,\ell} \cap (\bigcup_{j=0}^{k-1} V_j) = \emptyset\}$. Let $\Omega_0 = \{1\}$. We now put

$$(3.8) \quad R_n = \{\bigcup_{k=0}^{4^m} V_k\} \cup Q_{(4^m+1)m}, \quad f_n(z) = \prod_{k=0}^{4^m} \prod_{\ell \in \Omega_k} u(4^{km}(z - \zeta(Q_{km,\ell}))),$$

where

$$u(z) = \exp[e^{i\pi/4}\, 4^{-m} \{ -\frac{1}{z-(e^{i\pi/4}\, 4^{-m}/\sqrt{2})} + \frac{1}{z-1-i+(e^{i\pi/4}\, 4^{-m}/\sqrt{2})}\}].$$

Then (R_n, f_n) satisfies the required conditions. (See [46].)

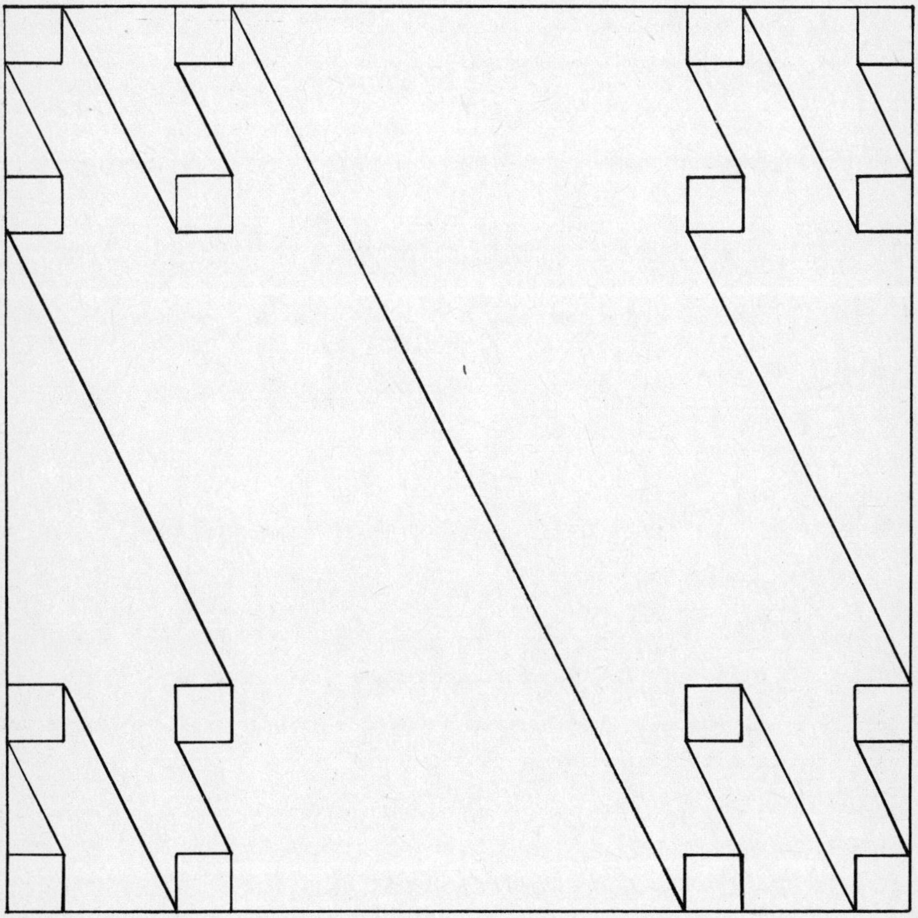

Calderón [5] suggests to study $C[a]$ for $a \notin L^\infty_{real}$. This problem seems very hard; in effect, Theorem D and Garnett's example immediately yield $\{\|C[a]\|_{2,2};\ a \in L^\infty_{real}\} = \infty$. (See Remark 3.16.) Let $Q'_n = \{(2-i)z/3;\ z \in Q_n\}$. (We rotate and contract Q_n.) Then Q'_n is a union of 4^n squares $\{Q'_{n,j}\}_{j=1}^{4^n}$ with sides of length $(\sqrt{5}/3)4^{-n}$; the sides of $Q'_{n,j}$ are not parallel to the coordinate axes. Let $I_{n,j}$ be the projection of $Q'_{n,j}$ to $\mathbb{R}$ and let $Q''_{n,j}$ be the segment in $Q'_{n,j}$ whose projection to $\mathbb{R}$ coinsides with $I_{n,j}$ $(1 \leq j \leq 4^n)$. The intervals $\{I_{n,j}\}_{j=1}^{4^n}$ are mutually non-overlapping, the length of each interval is 4^{-n} and their union equals $[0,1]$. Let $\Gamma_n = \bigcup_{j=1}^{4^n} Q'''_{n,j}$, where $Q'''_{n,j}$ is the closed sub-segment of $Q''_{n,j}$ of the same midpoint as $Q''_{n,j}$ such that $|Q'''_{n,j}| = |Q''_{n,j}|/2$. Then Γ_n is a locally chord-arc compact curve with constant 1. Since $|\Gamma_n| = \sqrt{10}/6$ and $\gamma(Q'_n) = (\sqrt{5}/3)\gamma(Q_n)$, Theorem D shows that

$$(3.9)\quad \|H_{\Gamma_n}\|_{L^1(\Gamma_n),L^1_w(\Gamma_n)} \geq \mathrm{Const}/\rho^*(\Gamma_n) \geq \mathrm{Const}/\rho(\Gamma_n)$$

$$\geq \mathrm{Const}\ |\Gamma_n|/\gamma(\Gamma_n) = \mathrm{Const}/\gamma(\Gamma_n)$$

$$\geq \mathrm{Const}/\gamma(Q'_n) = \mathrm{Const}/\gamma(Q_n).$$

We see that

$$\|H_{\Gamma_n}\|_{L^1(\Gamma_n),L^1_w(\Gamma_n)} \leq \mathrm{Const}\ \{\|H_{\Gamma_n}\|_{L^2(\Gamma_n),L^2(\Gamma_n)} + 1\}.$$

(See (3.18).) Since the projections of $\{Q'''_{n,j}\}_{j=1}^{4^n}$ to $\mathbb{R}$ are mutually disjoint, we can define a graph $\{(x, A_n(x));\ x \in \mathbb{R}\}$ containing Γ_n such that $A'_n\ (= a_n) \in L^\infty_{real}$. Since $a_n(x) = 1/3$ a.e. on the projection of Γ_n to $\mathbb{R}$,

$$\|H_{\Gamma_n}\|_{L^2(\Gamma_n),L^2(\Gamma_n)} \leq \mathrm{Const}\ \|C[a_n]\|_{2,2}.$$

Thus (3.9) shows that

$$1/\gamma(Q_n) \leq \mathrm{Const}\{\|C[a_n]\|_{2,2} + 1\}.$$

Since $\lim_{n\to\infty}\gamma(Q_n) = \gamma(Q_\infty) = 0$, this gives that $\{\|C[a]\|_{2,2};\ a \in L^\infty_{real}\} = \infty$.

It is very important to give various reasonable grounds to Vitushkin-Garnett's examples. From this point of view, we consider the following extremal problem. Let $I_0 = [0,1)$. For $s_1, \ldots, s_n \in \mathbb{R}$, we define

$$T_{s_1,\ldots,s_n}(x,y) = 1/\{(x-y) + i(A_{s_1,\ldots,s_n}(x) - A_{s_1,\ldots,s_n}(y))\},$$

where

$$A_{s_1,\ldots,s_n}(x) = \begin{cases} 0 & (x \notin I_0) \\ s_k & (\frac{k-1}{n} \leqq x < \frac{k}{n},\ 1 \leqq k \leqq n). \end{cases}$$

Our extremal problem is the following:

$$ex_\sigma(n) = \max\{\sigma(T_{s_1,\ldots,s_n});\ s_1,\ \ldots,\ s_n \in \mathbb{R}\} \qquad (n \geqq 1).$$

We see that

$$\text{Const}\sqrt{\log(n+1)} \leqq ex_\sigma(n) \leqq \text{Const}\sqrt{\log(n+1)} \qquad (n \geqq 1).$$

(See Appendix I.) We define a function A_n^0 on $\mathbb{R}$ by $A_n^0(x) = 0$ $(x \notin I_0)$ and $A_n^0(x) = \Sigma_{k=1}^{10^m} \varepsilon_k(x)$ $(x \in I_0)$, where m is the integral part of $(\log n)/(\log 10)$ and

$$\varepsilon_k(x) = \begin{cases} 0 & (j-1)10^{-k} \leqq x < j\ 10^{-k},\ 1 \leqq j \leqq 10^k\ ,\ j \text{ is odd} \\ 10^{-k} & j \text{ is even.} \end{cases}$$

Let $T_{A_n^0}(x,y)$ be the kernel associated with A_n^0. Then

$$\sigma(T_{A_n^0}) \geqq \text{Const}\sqrt{m} = \text{Const}\sqrt{\log(10^m)}.$$

(See (3.14).) Hence $\Gamma_n^0 = \{(x, A_n^0(x));\ x \in I_0\}$ is one of the worst graphs with respect to $ex_\sigma(n)$. The graph Γ_n^0 is similar to Q_m. Hence Theorem D suggests that

<u>Problem 3.2</u>. $$\text{Const}\ \gamma(\Gamma_n^0) \leqq \min\{\gamma(\Gamma_{s_1,\ldots,s_n});\ s_1,\ \ldots,\ s_n \in \mathbb{R}\}$$
$$\leqq \text{Const}\ \gamma(\Gamma_n^0) \qquad (n \geqq 1),$$

where $\Gamma_{s_1,\ldots,s_n} = \{(x, A_{s_1,\ldots,s_n}(x));\ x \in I_0\}$.

§3.3. The Cauchy transform on cranks

As a first step of harmonic analysis on discontinuous graphs, it is natural to begin with worst graphs. We say that a set $E \subset \mathbb{C}$ is thick, if there exists $M > 0$ such that, for any $z \in E$, $r > 0$,

$$1/M \leqq |E \cap D(z,r)|/r \leqq M.$$

The 1-dimension Calderón-Zygmund decomposition is applicable to thick sets. Hence thick sets are also natural objects. From the point of view of §3.2 and "thick sets", we define (thick) cranks.

An interval I in $I_0 = [0,1)$ is called a dyadic interval if I is expressed in the form $I = [(j-1)2^{-\ell}, j2^{-\ell})$ with integers $\ell \geq 0$, $1 \leq j \leq 2^{\ell}$. A finite sequence $\mathcal{R} = \{I_k\}_{k=1}^{m}$ of mutually disjoint dyadic intervals is called a covering (of I_0) if $I_0 = \cup_{k=1}^{m} I_k$, For a positive integer q and two coverings $\mathcal{R}' = \{I_j'\}_{j=1}^{n}$, $\mathcal{R} = \{I_k\}_{k=1}^{m}$, we write by $\mathcal{R}' \prec_q \mathcal{R}$ if each I_j' is expressed as a union of at least 2^q elements of $\mathcal{R}$ of same length. A segment I_0 is called a (thick) crank of degree 0. For a positive integer n, a graph $\Gamma = \{(x, A_\Gamma(x)); x \in I_0\}$ is called a (thick) crank of degree n, if there exist n coverings $\mathcal{R}_1, \ldots, \mathcal{R}_n$ and n functions $A_1, \ldots, A_n$ on I_0 such that

(3.10) $I_0 \prec_{q_1} \mathcal{R}_1 \prec_{q_2} \cdots \prec_{q_n} \mathcal{R}_n$ for some n tuple $(q_1, \ldots, q_n)$ of positive integers,

(3.11) $A_\Gamma = A_1 + \ldots + A_n$,

(3.12) on each element I of $\mathcal{R}_k$, A_k is a constant and $|A_k(x)| \leq |I|$ $(1 \leq k \leq n)$.

For two positive integers n, q and two real numbers α, β less than or equal to 1, we define a crank

$$\Gamma(n,q,\alpha,\beta) = \{(x, A_{\Gamma(n,q,\alpha,\beta)}(x));\ x \in I_0\}$$

by

$$A_{\Gamma(n,q,\alpha,\beta)} = A_1^{(q,\alpha,\beta)} + \ldots + A_n^{(q,\alpha,\beta)},$$

$$A_k^{(q,\alpha,\beta)}(x) = \begin{cases} \alpha 2^{-qk} & (x \in [(j-1)2^{-qk}, j\, 2^{-qk}),\ j \text{ odd}) \\ \beta 2^{-qk} & (x \in [(j-1)2^{-qk}, j\, 2^{-qk}),\ j \text{ even}). \end{cases}$$

We show

Theorem E. Let Γ be a crank of degree n. Then

(3.13) $$\|\mathcal{H}_\Gamma\|_{L^2(\Gamma), L^2(\Gamma)} \leq \text{Const}\sqrt{n}.$$

There exists an absolute constant η_0 such that, if $|\alpha - \beta|^2 \geq \eta_0/q$, then

(3.14) $$\|\mathcal{H}_{\Gamma(n)}\|_{L^2(\Gamma(n)), L^2(\Gamma(n))} \geq \text{Const}\sqrt{n}$$

$(\Gamma(n) = \Gamma(n,q,\alpha,\beta),\ n \geq 1)$.

In this section, we give the proof of the first half of Theorem E. For a crank $\Gamma = \{(x, A_\Gamma(x));\ x \in I_0\}$, we put $A_\Gamma(x) = 0$ outside I_0 and define a kernel

$$T_\Gamma(x,y) = 1/\{(x-y) + i(A_\Gamma(x) - A_\Gamma(y))\} \qquad (x \neq y,\ x, y \in \mathbb{R}).$$

For $f \in L^2(\Gamma)$, we have

$$\mathcal{H}_\Gamma f(x + i\, A_\Gamma(x))$$

$$= -\frac{1}{\pi}\ \text{p.v.} \int_0^1 T_\Gamma(x,y) f(y + i\, A_\Gamma(y))dy \qquad \text{a.e. on } I_0,$$

and hence

$$(3.15) \quad \|\mathcal{H}_\Gamma\|_{L^2(\Gamma),L^2(\Gamma)} \leqq \frac{1}{\pi} \|T_\Gamma\|_{2,2}.$$

Here are three lemmas necessary for the proof of (3.13).

Lemma 3.3. Let Γ be a crank of degree n. Then

$$\|T_\Gamma\|_{2,2} \leqq \text{Const}\ \{\sigma(T_\Gamma) + 1\}.$$

Proof. Evidently,

$$(3.16) \quad |T_\Gamma(x,y)| \leqq 1/|x-y|.$$

For any dyadic interval I, we have

$$(3.17) \quad |T_\Gamma(x,y) - T_\Gamma(x',y)| \leqq \text{Const}\ |I|/|x-y|^2$$
$$(x,\ x' \in I,\ y \notin I^* = (\text{the double of } I)).$$

This is shown as follows. Let $\mathcal{R}_1, \ldots, \mathcal{R}_n$ be the coverings associated with Γ and let $A_1, \ldots, A_n$ be the functions associated with A_Γ. We denote by m_I the smallest integer of k $(1 \leqq k \leqq n)$ such that $\mathcal{R}_k$ has an element contained in I. Then, for any $x, x' \in I$,

$$|A_k(x) - A_k(x')| \leqq \begin{cases} 0 & (1 \leqq k < m_I) \\ 2^{m_I - k + 1} |I| & (m_I \leqq k \leqq n). \end{cases}$$

Thus

$$|T_\Gamma(x,y) - T_\Gamma(x',y)| \leqq \text{Const}|(x + i\, A_\Gamma(x)) - (x' + i\, A_\Gamma(x'))|\ /|x-y|^2$$

$$\leqq \text{Const}\{|x-x'| + \sum_{k=m_I}^{n} |A_k(x) - A_k(x')|\}\ /|x-y|^2 \leqq \text{Const}|I|/|x-y|^2.$$

The proof of this lemma is analogous to the proof of Lemma 2.5; (3.16) and (3.17) play the same role as $\omega_1(\cdot)$. In the same manner as in the proof of (2.9), we obtain

$$\sigma(T_\Gamma^*) \leq \text{Const}\,\{\sigma(T_\Gamma) + 1\}.$$

(Replace $J' = (x - (\varepsilon/2), x + (\varepsilon/2))$ by the largest dyadic interval in J' containing x.) Using this inequality, we obtain

$$(3.18) \qquad |x;\ T_\Gamma^* f(x) > 3\lambda,\ Mf(x) \leq \eta\lambda|$$
$$\leq \frac{1}{100}\,|x;\ T_\Gamma^* f(x) > \lambda| \qquad (f \in L^2,\ \lambda > 0),$$

where $\eta = C_0\,\{\sigma(T_\Gamma) + 1\}$ and C_0 is a suitable constant. (Replace $\{I_k\}_{k=1}^{\infty}$ by a suitable sequence of dyadic intervals.) Inequality (3.18) immediately yields $\|T_\Gamma\|_{2,2} \leq \text{Const}\,\{\sigma(T_\Gamma) + 1\}$. Q.E.D.

For a non-negative integer n, we put

$$\sigma(n) = \sup\,\{\sigma(T_\Gamma);\ \Gamma \text{ crank of degree } \leq n\},$$
$$\hat{\tau}(n) = \sup\,\{\hat{\sigma}(I_0, T_\Gamma, f);\ f \in L^\infty_{\text{real}},\ 0 \leq f \leq 1,$$
$$\Gamma \text{ crank of degree } \leq n\}.$$

We show

Lemma 3.4. For two positive integers n, ℓ with $\ell \leq n-1$,

$$(3.19) \qquad \hat{\tau}(n) \leq \hat{\tau}(\ell) + \hat{\tau}(n - \ell - 1) + \text{Const}\,\sigma(n).$$

Proof. Let $f \in L^\infty_{\text{real}}$, $0 \leq f \leq 1$. Without loss of generality we may assume that $\text{supp}(f) \subset I_0$. Let Γ be a crank of degree n, $\mathcal{R}_1, \ldots, \mathcal{R}_n$ be n coverings satisfying (3.10) and let $A_1, \ldots, A_n$ be n functions satisfying (3.11), (3.12). Put

$$\Gamma' = \{(x, A_{\Gamma'}(x));\ x \in I_0\},\quad A_{\Gamma'} = A_1 + \ldots + A_{n-\ell-1},$$
$$\mathcal{R}_{n-\ell} = \{I_k\}_{k=1}^{m}.$$

Then Γ' is a crank of degree $n - \ell - 1$. We have

$$\hat{\sigma}(I_0, T_\Gamma, f) = (T_\Gamma f, T_\Gamma f)_{fdx} = \hat{\sigma}(I_0, T_{\Gamma'}, f)$$
$$+ ((T_\Gamma - T_{\Gamma'})f, T_\Gamma f)_{fdx} + (T_{\Gamma'} f, (T_\Gamma - T_{\Gamma'})f)_{fdx}$$

and

$$((T_\Gamma - T_{\Gamma'})f, T_\Gamma f)_{fdx}$$

$$= \sum_{k=1}^{m} \int_{I_k} (T_\Gamma - T_{\Gamma'})(\chi_{I_k} f)(x)\, \overline{T_\Gamma(\chi_{I_k} f)(x)}\, f(x)dx$$

$$+ \sum_{k=1}^{m} \int_{I_k} (T_\Gamma - T_{\Gamma'})(\chi_{I_k} f)(x)\, \overline{T_\Gamma(\chi_{I_k^c} f)(x)}\, f(x)dx$$

$$+ \sum_{k=1}^{m} \int_{I_k^c} (T_\Gamma - T_{\Gamma'})(\chi_{I_k} f)(x)\, \overline{T_\Gamma f(x)}\, f(x)dx$$

$$(= L_1 + L_2 + L_3, \quad \text{say}).$$

Since $T_{\Gamma'}(x,y) = 1/(x-y)$ $(x,y \in I_k,\ 1 \leqq k \leqq m)$ and $\sigma(n) \geqq \text{Const}$, we have, by Lemma 3.3,

$$|L_1| \leqq \sum_{k=1}^{m} \hat{\sigma}(I_k, T_\Gamma, f) + \pi \sum_{k=1}^{m} \int_{I_k} |H(\chi_{I_k} f)(x)\, T_\Gamma(\chi_{I_k} f)(x)|dx$$

$$\leqq \sum_{k=1}^{m} \hat{\sigma}(I_k, T_\Gamma, f) + \text{Const}\, \|T_\Gamma\|_{2,2}$$

$$\leqq \sum_{k=1}^{m} \hat{\sigma}(I_k, T_\Gamma, f) + \text{Const}\, \sigma(n).$$

Extending coordinates, we see that, for each I_k, there exist $f_k \in L^\infty_{\text{real}}$, $0 \leqq f_k \leqq 1$ and a crank Γ_k of degree ℓ such that

$$\hat{\sigma}(I_k, T_\Gamma, f) = |I_k|\, \hat{\sigma}(I_0, T_{\Gamma_k}, f_k).$$

Hence

$$|L_1| \leqq \sum_{k=1}^{m} |I_k|\, \hat{\sigma}(I_0, T_{\Gamma_k}, f_k) + \text{Const}\, \sigma(n)$$

$$\leqq \hat{\tau}(\ell) + \text{Const}\, \sigma(n).$$

Recall (3.16) and (3.17). Since $T_\Gamma(x,y) - T_{\Gamma'}(x,y)$ is anti-symmetric, we have, with x_k = (the midpoint of I_k), I_k^* = (the double of I_k),

$$|L_2| = \Big|\sum_{k=1}^{m} \int_{I_k} (T_\Gamma - T_{\Gamma'})(\chi_{I_k} f)(x)$$

$$\times \overline{\{T_\Gamma(\chi_{I_k^{*c}} f)(x) - T_\Gamma(\chi_{I_k^{*c}} f)(x_k)\}}\, f(x)\, dx$$

$$+ \sum_{k=1}^{m} \int_{I_k} (T_\Gamma - T_{\Gamma'})(\chi_{I_k} f)(x) \overline{T_\Gamma(\chi_{I_k^*-I_k} f)(x)} f(x)dx \Big|$$

$$\leq \text{Const} \sum_{k=1}^{m} \int_{I_k} |(T_\Gamma - T_{\Gamma'})(\chi_{I_k} f)(x)| \, Mf(x) \, dx$$

$$+ \sum_{k=1}^{m} \int_{I_k} |(T_\Gamma - T_{\Gamma'})(\chi_{I_k} f)(x)| \, (\int_{I_k^*-I_k} \frac{dy}{|x-y|}) \, dx$$

$$\leq \text{Const} \ \|T_\Gamma - T_{\Gamma'}\|_{2,2} \leq \text{Const} \ \sigma(n).$$

We have

$$|L_3| \leq \sum_{j,k;j\neq k} \int_{I_j} \{\int_{I_k} |T_\Gamma(x,y) - T_{\Gamma'}(x,y)|dy\} \, |T_\Gamma f(x)| \, dx$$

$$\leq \sum_{j,k;j\neq k} \int_{I_k^{*c} \cap I_j} \{\int_{I_j^{*c} \cap I_k} |T_\Gamma(x,y) - T_{\Gamma'}(x,y)|dy\} |T_\Gamma f(x)| \, dx$$

$$+ \sum_{j,k;j\neq k} \int_{I_k^{*c} \cap I_j} \{\int_{(I_j^{*c}-I_j) \cap I_k} | \, T_\Gamma(x,y)-T_{\Gamma'}(x,y)|dy\} \, |T_\Gamma f(x)|dx$$

$$+ \sum_{j,k;j\neq k} \int_{(I_k^*-I_k) \cap I_j} \{\int_{I_k} |T_\Gamma(x,y) - T_{\Gamma'}(x,y)|dy\} \, |T_\Gamma f(x)|dx$$

$$= L_{31} + L_{32} + L_{33},$$

$$|L_{33}| \leq \sum_{k=1}^{m} \int_{I_k^*-I_k} \{\int_{I_k} |T_\Gamma(x,y) - T_{\Gamma'}(x,y)|dy\} \, |T_\Gamma f(x)|dx$$

$$\leq 2 \sum_{k=1}^{m} \int_{I_k^*-I_k} (\int_{I_k} \frac{dy}{|x-y|}) \, |T_\Gamma f(x)|dx$$

$$\leq 2 \sum_{k=1}^{m} \{\int_{I_k^*-I_k} (\int_{I_k} \frac{dy}{|x-y|})^3 \, dx\}^{1/3} \{\int_{I_k^*-I_k} |T_\Gamma f(x)|^{3/2} \, dx\}^{2/3}$$

$$\leq \text{Const} \sum_{k=1}^{m} |I_k|^{1/3} \{\int_{I_k^*} |T_\Gamma f(x)|^{3/2} \, dx\}^{2/3}$$

$$\leq \text{Const} \ (\sum_{k=1}^{m} |I_k|)^{1/3} \{ \sum_{k=1}^{m} \int_{I_k^*} |T_\Gamma f(x)|^{3/2} \, dx\}^{2/3}$$

$$\leq \text{Const} \ \{ \int_{-\infty}^{\infty} (\sum_{k=1}^{m} \frac{|I_k|^2}{(x-x_k)^2 + |I_k|^2}) \, |T_\Gamma f(x)|^{3/2} \, dx\}^{2/3}$$

$$\leq \text{Const } \{\int_{-\infty}^{\infty} (\sum_{k=1}^{m} \frac{|I_k|^2}{(x-x_k)^2 + |I_k|^2})^4\}^{1/6} \{\int_{-\infty}^{\infty} |T_\Gamma f(x)|^2 \, dx\}^{1/2}$$

$$\leq \text{Const } \|T_\Gamma f\|_2 \leq \text{Const } \|T_\Gamma\|_{2,2} \leq \text{Const } \sigma(n),$$

$$|L_{3,2}| \leq \sum_{j=1}^{m} \int_{I_j} \{\int_{I_j^*-I_j} |T_\Gamma(x,y) - T_{\Gamma'}(x,y)|dy\} |T_\Gamma f(x)|dx$$

$$\leq 2 \sum_{j=1}^{m} \int_{I_j} (\int_{I_j^*-I_j} \frac{dy}{|x-y|}) |T_\Gamma f(x)|dx$$

$$\leq \text{Const } \|T_\Gamma f\|_2 \leq \text{Const } \sigma(n)$$

and

$$|L_{31}|$$

$$\leq \sum_{j,k;j\neq k} \int_{I_k^{*c} \cap I_j} \{\int_{I_j^{*c} \cap I_k} \frac{|A_\Gamma(x) - A_{\Gamma'}(x)| + |A_\Gamma(y) - A_{\Gamma'}(y)|}{|x - y|^2} dy\} |T_\Gamma f(x)|dx$$

$$\leq \sum_{j=1}^{m} \int_{I_j} |\sum_{\mu=n-\ell}^{n} A_\mu(x)| (\int_{I_j^{*c}} \frac{dy}{|x-y|^2}) |T_\Gamma f(x)|dx$$

$$+ \sum_{k=1}^{m} \int_{I_k^{*c}} \{\int_{I_k} |\sum_{\mu=n-\ell}^{n} A_\mu(y)|/|x-y|^2 \, dy\} |T_\Gamma f(x)|dx$$

$$\leq \text{Const } \sum_{j=1}^{m} \int_{I_j} |I_j| (\int_{I_j^{*c}} \frac{dy}{|x-y|^2}) |T_\Gamma f(x)|dx$$

$$+ \text{Const } \sum_{k=1}^{m} \int_{I_k^{*c}} |I_k| (\int_{I_k} \frac{dy}{|x-y|^2}) |T_\Gamma f(x)|dx$$

$$\leq \text{Const } \int_{I_0} |T_\Gamma f(x)|dx + \text{Const } \int_{-\infty}^{\infty} (\sum_{k=1}^{m} \frac{|I_k|^2}{|x-x_k|^2 + |I_k|^2}) |T_\Gamma f(x)|dx$$

$$\leq \text{Const } \|T_\Gamma f\|_2 \leq \text{Const } \sigma(n).$$

Thus

$$|L_3| \leq \text{Const } \sigma(n).$$

Consequently,

$$|((T_\Gamma - T_{\Gamma'})f, T_\Gamma f)_{fdx}| \leq |L_1| + |L_2| + |L_3| \leq \hat{\tau}(\ell) + \text{Const } \sigma(n).$$

In the same manner, we have

$$|(T_{\Gamma'}f,\ (T_\Gamma - T_{\Gamma'})f)_{fdx}|$$

$$\le \Big|\sum_{k=1}^{m} \int_{I_k} T_{\Gamma'}(\chi_{I_k} f)(x)\ \overline{(T_\Gamma - T_{\Gamma'})(\chi_{I_k} f)(x)}\ f(x)dx\Big| + \text{Const}\ \sigma(n).$$

Since $T_{\Gamma'}(x,y) = 1/(x-y)$ $(x,y \in I_k,\ 1 \le k \le m)$, the first quantity in the right-hand side is dominated by Const $\sigma(n)$. Thus

$$\hat\sigma(I_0,\ T_\Gamma,\ f) \le \hat\sigma(I_0,\ T_{\Gamma'},\ f) + \hat\tau(\ell) + \text{Const}\ \sigma(n)$$

$$\le \hat\tau(n - \ell - 1) + \hat\tau(\ell) + \text{Const}\ \sigma(n),$$

which gives (3.19). Q.E.D.

Lemma 3.5. $\sigma(n)^2 \leqq \text{Const}\ \hat\tau(n)$ $(n \geqq 0)$.

Proof. We see that, for any crank Γ, $\sigma(T_\Gamma)^2 \le \text{Const}\ \hat\sigma(T_\Gamma)$. (See (Second Step) in §2.8.) Hence it is sufficent to show that

(3.20) $\quad \sup\ \{\hat\sigma(T_\Gamma);\ \Gamma$ crank of degree $n\} \le \text{Const}\ \hat\tau(n)$.

Since $\hat\tau(0) \geqq$ Const, (3.20) holds for $n = 0$. Let Γ be a crank of degree $n \geqq 1$ and let $f \in L^\infty_{\text{real}}$, $0 \le f \le 1$. For any dyadic interval I, there exist a crank Γ_I of degree $\le n$ and $f_I \in L^\infty_{\text{real}}$, $0 \le f_I \le 1$ such that $\hat\sigma(I,\ T_\Gamma,\ f) = |I|\ \hat\sigma(I_0,\ T_{\Gamma_I},f_I)$. Hence $(1/|I|)\ \hat\sigma(I,\ T_\Gamma,\ f) \le \hat\tau(n)$. For any non-dyadic interval $I \subset I_0$, there exist mutually disjoint two dyadic intervals I_1, I_2 such that $|I_1| = |I_2| \le 2|I|$, $I \subset I_1 \cup I_2$. Then

$$\frac{1}{|I|}\ \hat\sigma(I,\ T_\Gamma,\ f) = \frac{1}{|I|}\ \hat\sigma(I_1 \cup I_2,\ T_\Gamma,\ \chi_I f)$$

$$\le \frac{2}{|I|}\ \{\hat\sigma(I_1,\ T_\Gamma,\ \chi_I f) + \hat\sigma(I_2,\ T_\Gamma,\ \chi_I f)$$

$$+ \int_{I_1} \Big(\int_{I_2} \frac{dy}{|x-y|}\Big)^2 dx + \int_{I_2}\Big(\int_{I_1} \frac{dy}{|x-y|}\Big)^2 dx\}$$

$$\le \text{Const}\ \{\hat\tau(n) + 1\} \le \text{Const}\ \hat\tau(n).$$

For any interval $I \subset \mathbb{R}$, we have

$$\frac{1}{|I|}\ \hat\sigma(I,\ T_\Gamma,f) \le \frac{1}{|I|}\ \hat\sigma(I_0 \cap I,\ T_\Gamma,f) + \text{Const} \le \text{Const}\ \hat\tau(n).$$

Thus $\hat\sigma(T_\Gamma) \le \text{Const}\ \hat\tau(n)$, which gives (3.20). Q.E.D.

In the same manner as in Lemmas 3.4 and 3.5, we have

Lemma 3.6. $\sigma(n) \leqq \sigma(n-1) + \mathrm{Const} \quad (n \geqq 1)$.

We now give the proof of (3.13). Since $\sigma(0) \leqq \mathrm{Const}$, Lemmas 3.4 and 3.6 show that $\hat{\tau}(n) < \infty$ for all $n \geqq 1$. By Lemmas 3.4 and 3.5, we have, for any non-negative integer m,

$$\hat{\tau}(2^m) \leqq \hat{\tau}(2^{m-1}) + \hat{\tau}(2^{m-1} - 1) + \mathrm{Const}\ \sigma(2^m)$$

$$\leqq 2\ \hat{\tau}(2^{m-1}) + \mathrm{Const}\ \sigma(2^m) \leqq \dots$$

$$\leqq 2^m\ \hat{\tau}(1) + \mathrm{Const} \sum_{k=1}^{m} 2^{m-k}\ \sigma(2^k) \leqq 2^m\ \hat{\tau}(1) + \mathrm{Const}\ 2^m\ \sigma(2^m)$$

$$\leqq \mathrm{Const}\ 2^m\ \hat{\tau}(2^m)^{1/2},$$

which shows that $\hat{\tau}(2^m) \leqq \mathrm{Const}\ 2^{2m}$. Hence

$$\hat{\tau}(2^m) \leqq 2^m\ \hat{\tau}(1) + \mathrm{Const} \sum_{k=1}^{m} 2^{m-k}\ \hat{\tau}(k)^{1/2}$$

$$\leqq 2^m\ \hat{\tau}(1) + \mathrm{Const} \sum_{k=1}^{m} 2^{m-k}\ 2^k \leqq \mathrm{Const}(m+1)2^m.$$

Consequently,

$$\hat{\tau}(2^m) \leqq 2^m\ \hat{\tau}(1) + \mathrm{Const} \sum_{k=1}^{m} 2^{m-k}\ \hat{\tau}(k)^{1/2}$$

$$\leqq 2^m\ \hat{\tau}(1) + \mathrm{Const} \sum_{k=1}^{m} 2^{m-k} \sqrt{k+1}\ 2^{k/2} \leqq \mathrm{Const}\ 2^m.$$

For an integer $n \geqq 1$, we choose a non-negative integer m so that $2^m \leqq n < 2^{m+1}$. Then

$$\hat{\tau}(n) \leqq \hat{\tau}(2^{m+1}) \leqq \mathrm{Const}\ 2^{m+1} \leqq \mathrm{Const}\ n.$$

Consequently, Lemmas 3.3 and 3.5 yield (3.13).

§3.4. Proof of the latter half of Theorem E

The following idea is essentially due to David [18, Chap. III]. Fixing q, α, β, we write

$$\Gamma(n) = \Gamma(n, q, \alpha, \beta), \quad A_n = A_n^{(q,\alpha,\beta)} \quad (n \geqq 1).$$

We put

$$T_n^0(x,y) = T_{\Gamma(n)}(x,y) - \frac{1}{x-y},$$

$$\tilde{\tau}_0(n) = \tilde{\sigma}(I_0, T_n^0, \chi_{I_0}) \quad (n \geqq 1).$$

Here are two lemmas necessary for the proof.

Lemma 3.7. $\tilde{\tau}_0(1) \geqq |\alpha - \beta|^2/100$.

Proof. Let $x \in [(k_0-1)2^{-q}, k_0 2^{-q})$ (k_0 is odd). Then

$$| T_1^0 \chi_{I_0}(x) | = |\int_{I_0} \{ \frac{1}{(x-y) + i(A_1(x)-A_1(y))} - \frac{1}{x-y}\} dy|$$

$$= |\alpha - \beta| \, 2^{-q} \Big| \sum_{k \text{ even}} \int_{[(k-1)2^{-q},k2^{-q})} \frac{dy}{\{(x-y) + i(\alpha-\beta)2^{-q}\}(x-y)} \Big|$$

$$\geqq |\alpha - \beta| 2^{-q} \Big| \mathrm{Re} \sum_{k \text{ even}} \int_{[(k-1)2^{-q},k2^{-q})} \Big|$$

$$= |\alpha - \beta| 2^{-q} \sum_{k \text{ even}} \int_{[(k-1)2^{-q},k2^{-q})} \frac{dy}{(x-y)^2 + (\alpha-\beta)^2 \, 2^{-2q}}$$

$$\geqq |\alpha - \beta| 2^{-q} \int_{2^{-q}}^{2^{-q+1}} \frac{dy}{y^2 + (\alpha-\beta)^2 \, 2^{-2q}} \geqq |\alpha - \beta|/10.$$

In the same manner, we have, for any $x \in [(k_0-1)2^{-q}, k_0 2^{-q})$ (k_0 is even),

$$|T_1^0 \chi_{I_0}(x)| \geqq |\alpha - \beta|/10.$$

Thus

$$\tilde{\tau}_0(1) = \tilde{\sigma}(I_0, T_1^0, \chi_{I_0}) \geqq \int_{I_0} \left(\frac{|\alpha-\beta|}{10} \right)^2 dx \geqq |\alpha - \beta|^2/100. \qquad \text{Q.E.D.}$$

Lemma 3.8. For two positive integers n, ℓ with $\ell \leqq n - 1$,

$$\tilde{\tau}_0(n) \geqq \tilde{\tau}_0(\ell) + \tilde{\tau}_0(n - \ell) - \text{Const } q^{-1}\sqrt{n}. \tag{3.21}$$

Proof. We write

$$I_k = [(k-1)2^{-q\ell}, k2^{-q\ell}) \quad (1 \leqq k \leqq 2^{q\ell}).$$

We have

$$\tilde{\tau}_0(n) = \int_{I_0} |T_n^0 \chi_{I_0}(x)|^2 dx = \tilde{\tau}_0(\ell)$$

$$+ \int_{I_0} (T_n^0 - T_\ell^0) \chi_{I_0}(x) \overline{T_n^0 \chi_{I_0}(x)} dx + \int_{I_0} T_\ell^0 \chi_{I_0}(x) \overline{(T_n^0 - T_\ell^0) \chi_{I_0}(x)} dx$$

$$= \tilde{\tau}_0(\ell) + L_1 + L_2,$$

$$L_1 = \sum_{k=1}^{2^{q\ell}} \int_{I_k} \{\int_{I_k} (T_n^0(x,y) - T_\ell^0(x,y))dy\}\{\overline{\int_{I_k} T_n^0(x,y)dy}\} dx$$

$$+ \sum_{k=1}^{2^{q\ell}} \int_{I_k} \{\int_{I_k} (T_n^0(x,y) - T_\ell^0(x,y))dy\} \overline{\{\int_{I_0-I_k} T_n^0(x,y)dy\}}dx$$

$$+ \sum_{k=1}^{2^{q\ell}} \int_{I_0-I_k} \{\int_{I_k} (T_n^0(x,y) - T_\ell^0(x,y))dy\} \overline{T_n^0 \chi_{I_0}(x)}\, dx$$

$$= L_{11} + L_{12} + L_{13}.$$

Note that $T_\ell^0(x,y) = 0 \quad (x,\ y \in I_k,\ 1 \leq k \leq 2^{q\ell})$ and

$$T_n^0(x,y) = [(x-y) + i \sum_{\mu = \ell+1}^{n} (A_\mu(x) - A_\mu(y))]^{-1} - \frac{1}{x-y}$$
$$(x,\ y \in I_k,\ 1 \leq k \leq 2^{q\ell}) .$$

Hence, extending the coordinate axes, we have

$$L_{11} = \sum_{k=1}^{2^{q\ell}} \int_{I_k} |\int_{I_k} T_n^0(x,y)dy|^2\, dx = \sum_{k=1}^{2^{q\ell}} |I_k|\ \tilde{\tau}_0(n - \ell) = \tilde{\tau}_0(n - \ell).$$

Let p be the integer such that $q^4 < p \leq 2q^4$ and $(\log p)/\log 2$ is an integer. For each $1 \leq k \leq 2^{q\ell}$, we write $I_k = I_{k,1} \cup \ldots \cup I_{k,p^2}$, where $\{I_{k,j}\}_{j=1}^{p^2}$ are mutually disjoint dyadic intervals of length $p^{-2}\, 2^{q\ell}$. Let $\tilde{I}_k$ denote the closed interval of the same midpoint as I_k and of length $(1 + p^{-1})\ |I_k|$, and let $\tilde{\tilde{I}}_{k,j}$ denote the closed interval of the same midpoint as $I_{k,j}$ and of length $(1 + p^{-4})\ |I_{k,j}|$. We have, with x_k = (the midpoint of I_k), $x_{k,j}$ = (the midpoint of $I_{k,j}$),

$$L_{12} = \sum_{k=1}^{2^{q\ell}} \int_{I_k} \{\int_{I_k} T_n^0(x,y)dy\} \overline{\{\int_{I_0-I_k} T_n^0(x,y)dy\}}\, dx$$

$$= \sum_{k=1}^{2^{q\ell}} \int_{I_k} \{\int_{I_k} T_n^0(x,y)dy\} \overline{\{\int_{(I_0 \cap \tilde{I}_k)-I_k} T_n^0(x,y)dy\}}dx$$

$$+ \sum_{k=1}^{2^{q\ell}} \int_{I_k} \{\int_{I_k} T_n^0(x,y)dy\} \overline{\{\int_{I_0-(I_0 \cap \tilde{I}_k)} (T_n^0(x,y)-T_n^0(x_k,y))dy\}}dx$$

$$= L_{121} + L_{122} ,$$

$$|L_{121}| \leq 2 \sum_{k=1}^{2^{q\ell}} \int_{I_k} |\int_{I_k} T_n^0(x,y)dy|\ \left(\int_{\tilde{I}_k-I_k} \frac{dy}{|x-y|}\right) dx$$

$$\leq 2 \{ \sum_{k=1}^{2^{q\ell}} \tilde{\sigma}(I_k, T_n^0, \chi_{I_k}) \}^{1/2} \{ \sum_{k=1}^{2^{q\ell}} \int_{I_k} (\int_{\tilde{I}_k - I_k} \frac{dy}{|x-y|})^2 \, dx\}^{1/2}$$

$$\leq \text{Const} \ \{\int_0^1 \log^2 (1 + \frac{1}{ps}) ds\}^{1/2} \ \tilde{\tau}_0(n - \ell)^{1/2}$$

$$\leq \text{Const} \ p^{-1/2} \ \tilde{\tau}_0(n - \ell)^{1/2}$$

and

$$L_{122} = \sum_{k=1}^{2^{q\ell}} \sum_{j=1}^{p^2} \int_{I_{k,j}} \{\int_{I_{k,j}} T_n^0(x,y)\,dy\} \overline{\{\int_{I_0-(I_0 \cap \tilde{I}_k)} (T_n^0(x,y) - T_n^0(x_k,y))\,dy\}}\, dx$$

$$+ \sum_{k=1}^{2^{q\ell}} \sum_{j=1}^{p^2} \int_{(I_k \cap \tilde{\tilde{I}}_{k,j}) - I_{k,j}} \{\int_{I_{k,j}} \} \overline{\{\int_{I_0-(I_0 \cap \tilde{I}_k)}\}}$$

$$+ \sum_{k=1}^{2^{q\ell}} \sum_{j=1}^{p^2} \int_{I_k \cap \tilde{\tilde{I}}_{k,j}^c} \{\int_{I_{k,j}} \} \overline{\{\int_{I_0-(I_0 \cap \tilde{I}_k)}\}}$$

$$= L_{1221} + L_{1222} + L_{1223} \, .$$

Since

$$|T_n^0(x,y) - T_n^0(x_{k,j}, y)| \leq \frac{2|x - x_{k,j}| + |\sum_{\mu=\ell+1}^{n} (A_\mu(x) - A_\mu(x_{k,j}))|}{|x-y| \ |x_{k,j} - y|}$$

$$\leq \text{Const} \ \{p^{-2} \, 2^{-q\ell} + 2^{-q(\ell+1)}\} \ / \ |x_{k,j} - y|^2$$

$$\leq \text{Const} \ p^{-2} \, 2^{-q\ell} / |x_{k,j} - y|^2 \qquad (x \in I_{k,j}, \ y \in \tilde{I}_k^c),$$

(3.13) shows that

$$|L_{1221}| = \Big| \sum_{k=1}^{2^{q\ell}} \sum_{j=1}^{p^2} \int_{I_{k,j}} \{\int_{I_{k,j}} T_n^0(x,y)\,dy\}$$

$$\times \overline{\{\int_{I_0-(I_0 \cap \tilde{I}_k)} (T_n^0(x,y) - T_n^0(x_{k,j},y))\,dy\}} \, dx \Big|$$

$$\leq \text{Const} \sum_{k=1}^{2^{q\ell}} \sum_{j=1}^{p^2} \int_{I_{k,j}} |\int_{I_{k,j}} T_n^0(x,y)\,dy| \ (\int_{\tilde{I}_k^c} \frac{p^{-2} \, 2^{-q\ell}}{|x_{k,j} - y|^2} \, dy) \, dx$$

$$\leq \text{Const} \ p^{-1} \sum_{k=1}^{2^{q\ell}} \sum_{j=1}^{p^2} \int_{I_{k,j}} |\int_{I_{k,j}} T_n^0(x,y)\,dy|$$

$$\leq \text{ Const } p^{-1} \|T_n^0\|_{2,2} \leq \text{ Const } p^{-1} \sqrt{n} .$$

Since

$$|T_n^0(x,y) - T_n^0(x_k,y)| \leq \text{ Const } p\, 2^{-q\ell}/|x_k-y|^2 \quad (x \in I_k,\ y \in \tilde{I}_k^c),$$

$$|T_n^0(x,y)| \leq \text{ Const } 2^{-q(\ell+1)}/|x-y|^2 \leq \text{ Const } 2^{-q(\ell+1)}\, p^8\, |I_{k,j}|^{-2}$$

$$\leq \text{ Const } p^{10}\, 2^{-q}/|I_{k,j}| \quad (x \in I_k \cap \tilde{\tilde{I}}_{k,j}^c,\ y \in I_{k,j})$$

we have

$$|L_{1222}| \leq \text{ Const } \sum_{k=1}^{2^{q\ell}} \sum_{j=1}^{p^2} \int_{\tilde{\tilde{I}}_{k,j}-I_{k,j}} \left(\int_{I_{k,j}} \frac{dy}{|x-y|}\right)\left(\int_{\tilde{I}_k^c} \frac{p\, 2^{-q\ell}}{|x_k-y|^2}\, dy\right)dx$$

$$\leq \text{ Const } p^2 \int_0^1 \log(1 + p^{-4}s^{-1})\, ds \leq \text{ Const}/p$$

and

$$|L_{1223}| \leq \text{ Const } \sum_{k=1}^{2^{q\ell}} \sum_{j=1}^{p^2} \int_{I_k \cap \tilde{\tilde{I}}_{k,j}^c} (\text{Const } p^{10} 2^{-q})\left(\int_{\tilde{I}_k^c} \frac{p\, 2^{-q\ell}}{|x_k-y|^2}\, dy\right) dx$$

$$\leq \text{ Const } p^{14}\, 2^{-q} \sum_{k=1}^{2^{q\ell}} \int_{I_k} dx \leq \text{ Const } p^{14}\, 2^{-q} .$$

Thus, by (3.13),

$$|L_{12}| \leq |L_{121}| + |L_{1221}| + |L_{1222}| + |L_{1223}|$$

$$\leq \text{ Const } \{p^{-1/2}\, \tilde{\tau}_0(n - \ell)^{1/2} + p^{-1}\sqrt{n} + p^{-1} + p^{14}\, 2^{-q}\}$$

$$\leq \text{ Const } p^{-1/2}\, \{\tilde{\tau}_0(n - \ell)^{1/2} + \sqrt{n}\} \leq \text{ Const } \sqrt{n}/q.$$

Since

$$|T_n^0(x,y) - T_\ell^0(x,y)| \leq \text{ Const } 2^{-q(\ell+1)}/|x-y|^2$$

$$\leq \text{ Const } p^2\, 2^{-q}\, |I_k|/\{|x-x_k|^2 + |I_k|^2\} \quad (x \in \tilde{I}_k^c,\ y \in I_k),$$

we have, in the same manner as in the estimate of $|L_{33}|$ in §3.3,

$$|L_{13}| \leq \text{ Const } \sum_{k=1}^{2^{q\ell}} \int_{(I_0 \cap \tilde{I}_k)-I_k} \left(\int_{I_k} \frac{dy}{|x-y|}\right) |T_n^0 \chi_{I_0}(x)|dx$$

$$+ \text{Const}\ p^2\, 2^{-q} \int_{I_0} \Big\{ \sum_{k=1}^{2^{q\ell}} \frac{|I_k|^2}{|x-x_k|^2 + |I_k|^2} \Big\}\, |T_n^0 \chi_{I_0}(x)|\, dx$$

$$\leqq \text{Const} \Big\{ \sum_{k=1}^{2^{q\ell}} \int_{\tilde{I}_k - I_k} \Big(\int_{I_k} \frac{dy}{|x-y|} \Big)^3 dx \Big\}^{1/3}$$

$$\times \Big\{ \sum_{k=1}^{2^{q\ell}} \int_{\tilde{I}_k} |T_n^0 \chi_{I_0}(x)|^{3/2} dx \Big\}^{2/3}$$

$$+ \text{Const}\ p^2\, 2^{-q} \Big\{ \int_{I_0} \Big(\sum_{k=1}^{2^{q\ell}} \frac{|I_k|^2}{|x-x_k|^2 + |I_k|^2} \Big)^2 dx \Big\}^{1/2} \Big\{ \int_{I_0} |T_n^0 \chi_{I_0}(x)|^2 dx \Big\}^{1/2}$$

$$\leqq \text{Const}\ \Big\{ \int_0^{1/p} \log^3 (1 + \tfrac{1}{s}) ds \Big\}^{1/3} \Big\{ \sum_{k=1}^{2^{q\ell}} \int_{\tilde{I}_k} |T_n^0 \chi_{I_0}(x)|^{3/2} dx \Big\}^{2/3}$$

$$+ \text{Const}\ p^2\, 2^{-q}\, \|T_n^0\|_{2,2}$$

$$\leqq \text{Const}\ (p^{-1/4} + p^2\, 2^{-q})\, \|T_n^0\|_{2,2} \leqq \text{Const}\ q^{-1} \sqrt{n}.$$

Thus

$$L_1 \geqq L_{11} - |L_{12}| - |L_{13}| \geqq \tilde{\tau}_0(n - \ell) - \text{Const}\ q^{-1} \sqrt{n}.$$

In the same manner,

$$|L_2| \leqq \Big| \sum_{k=1}^{2^{q\ell}} \int_{I_k} \Big\{ \int_{I_k} T_\ell^0(x,y) dy \Big\} \overline{\Big\{ \int_{I_k} (T_n^0(x,y) - T_\ell^0(x,y)) dy \Big\}}\, dx \Big|$$

$$+ \text{Const}\ q^{-1} \sqrt{n}.$$

Since $T_\ell^0(x,y) = 0$ $(x, y \in I_k,\ 1 \leqq k \leqq 2^{q\ell})$, $|L_2| \leqq \text{Const}\ q^{-1}\sqrt{n}$. Consequently,

$$\tilde{\tau}_0(n) \geqq \tilde{\tau}_0(\ell) + L_1 - |L_2| \geqq \tilde{\tau}_0(\ell) + \tilde{\tau}_0(n - \ell) - \text{Const}\ q^{-1} \sqrt{n}. \qquad \text{Q.E.D.}$$

We now give the proof of (3.14). Lemmas 3.7 and 3.8 show that, for any positive integer m,

$$\tilde{\tau}_0(2^m) \geqq 2\, \tilde{\tau}_0(2^{m-1}) - \text{Const}\ q^{-1}\, 2^{m/2} \geqq \ldots$$

$$\geqq 2^m\, \tilde{\tau}_0(1) - \text{Const}\ q^{-1} \sum_{k=0}^{m-1} 2^k\, 2^{(m-k)/2}$$

$$\geqq 2^m\, \{ |\alpha - \beta|^2/100 - \text{Const}/q \}\ .$$

For any integer $n \geqq 2$, we choose a positive integer m so that $2^{m-1} < n \leqq 2^m$. Then

$$\tilde{\tau}_0(n) \geqq \tilde{\tau}_0(2^{m-1}) + \tilde{\tau}_0(n - 2^{m-1}) - \text{Const } q^{-1}\sqrt{n}$$

$$\geqq 2^{m-1}\{|\alpha - \beta|^2/100 - \text{Const}/q\} - \text{Const } q^{-1}\sqrt{n}$$

$$\geqq \sqrt{n}\{|\alpha - \beta|^2/200 - \text{Const}/q\} .$$

Thus (3.14) holds if η_0 is large enough. This completes the proof of the latter half of Theorem E.

Corollary 3.9. For any crank Γ of degree $n \geqq 1$, $\rho(\Gamma) \geqq \rho_+(\Gamma) \geqq \text{Const}/\sqrt{n}$. If $|\alpha - \beta|^2 \geqq \eta_0/q$, then $\rho(\Gamma(n))^{1/3} \leqq \text{Const } \rho_+(\Gamma(n)) \leqq \text{Const}/\sqrt{n}$ ($\Gamma(n) = \Gamma(n,q,\alpha,\beta)$, $n \geqq 1$), where η_0 is the constant in Theorem E.

Proof. For any crank Γ of degree $n \geqq 1$, we have

$$\|H_\Gamma\|_{L^1(\Gamma),L^1_w(\Gamma)} \leqq \text{Const } \|T_\Gamma\|_{L^1(\mathbb{R}),L^1_w(\mathbb{R})} .$$

Since $\overline{\Gamma}$ is a locally chord-arc compact curve with constant 1, Theorem D is valid for Γ. Thus Theorem D, Lemma 3.3 and (3.13) yield the required inequalities. The latter half is also deduced from Theorem D, Lemma 3.3 and (3.14). Q.E.D.

David [18, Chap. III] showed that (2.39) is best possible in the following sense:

$$(3.22) \qquad \sup\{\|C[a]\|_{2,2};\ a \in L^\infty_{\text{real},M}\} \geqq \text{Const}\sqrt{M} \qquad (M \geqq 1).$$

This is deduced from Theorem E as follows. We showed that

$$\|T_{\Gamma_0(n)}\, \chi_{I_0}\|_2 \geqq \text{Const}\sqrt{n} \qquad (n \geqq 1) ,$$

where $\Gamma_0(n) = \Gamma(n, 1 + (\text{the integral part of } \eta_0), 1, 0)$. Adding some segments parallel to the y-axis to $\Gamma_0(n)$, we define an arc Λ_n with endpoints 0 and 1. Then $|\Lambda_n| \leqq \text{Const } n$. There exists a Lipschitz graph $\Lambda_n^* = \{(x, \int_0^x b_n^*(s)ds);\ x \in I_0\}$ such that $|\Lambda_n^*| \leqq C_0 n$ and $|\Gamma_0(n) \cup \Lambda_n^* - \Gamma_0(n) \cap \Lambda_n^*| \leqq \varepsilon$, where C_0 is an absolute constant and $0 < \varepsilon < 1/2$ is determined later. We have

$$\int_{I_0} |b_n^*(x)|dx \leqq |\Lambda_n^*| \leqq C_0 n.$$

Lemma 2.2 shows that there exists an open set $\Omega = \bigcup_{k=1}^{\infty} I_k$ $(I_k = I_{\Omega,k})$ in I_0 such that

$$|\Omega| \leqq \varepsilon, \quad (|b_n^*|)_{I_k} \leqq C_0 n/\varepsilon \quad (k \geqq 1),$$

$$|b_n^*(x)| \leqq C_0 n/\varepsilon \qquad \text{a.e. on } I_0 - \Omega .$$

We put

$$b_n^{**}(x) = \begin{cases} b_n^*(x) & (x \in I_0 - \Omega) \\ (b_n^*)_{I_k} & (x \in I_k,\ k \geqq 1) \end{cases}$$

and $\Lambda_n^{**} = \{(x, \int_0^x b_n^{**}(s)ds);\ x \in I_0\}$. Then $\|b_n^{**}\|_\infty \leqq C_0\, n/\varepsilon$. Let E be the projection of $\Gamma_0(n) \cap \Lambda_n^{**}$ to $\mathbb{R}$ and let $F = I_0 - E$. Since

$$|x \in I_0;\ \int_0^x b_n^*(s)ds \neq \int_0^x b_n^{**}(s)ds| = |\Omega| \leqq \varepsilon,$$

we have $|F| \leqq 2\varepsilon$. From the definition, $C[b_n^{**}](x,y) = T_{\Gamma_0(n)}(x,y)$ $(x,y \in E)$. Hence Theorem E shows that

$$\{\int_E |C[b_n^{**}]\,\chi_E(x)|^2 dx\}^{1/2} = \{\int_E |T_{\Gamma_0(n)}\,\chi_E(x)|^2\, dx\}^{1/2}$$

$$\geqq \{\int_E |T_{\Gamma_0(n)}\,\chi_{I_0}(x)|^2\, dx\}^{1/2} - \{\int_{I_0} |T_{\Gamma_0(n)}\,\chi_F(x)|^2\, dx\}^{1/2}$$

$$\geqq \{\text{Const } n - \int_F |T_{\Gamma_0(n)}\,\chi_{I_0}(x)|^2\, dx\}^{1/2} - \text{Const}\,\sqrt{n}\,\sqrt{\varepsilon}.$$

By (3.13) and (3.18), we have

$$\|T_{\Gamma_0(n)}\|_{4,4} \leqq \text{Const}\,\{\sigma(T_{\Gamma_0(n)}) + 1\} \leqq \text{Const}\,\sqrt{n},$$

and hence,

$$\int_F |T_{\Gamma_0(n)}\,\chi_{I_0}(x)|^2\, dx \leqq \|T_{\Gamma_0(n)}\|_{4,4}^2\, |F|^{1/2} \leqq \text{Const } n\,\sqrt{\varepsilon}.$$

Thus

$$\{\int_{I_0} |C[b_n^{**}]\,\chi_E(x)|^2\, dx\}^{1/2} \geqq \text{Const}\,\sqrt{n}\ \{(1 - \text{Const}\sqrt{\varepsilon}\,)^{1/2} - \text{Const}\,\sqrt{\varepsilon}\}.$$

Choosing ε sufficiently small, we have

$$b_n^{**} \in L^\infty_{\text{real},C_0 n/\varepsilon}, \quad \|C[b_n^{**}]\|_{2,2} \geqq \text{Const}\,\sqrt{n} \quad (n \geqq 1),$$

which yields (3.22).

§3.5. Analytic capacities of fat cranks

For $\rho > 0$, $z \in \mathbb{C}$ and $E \subset \mathbb{C}$, we write $[\rho E + z] = \{\rho\zeta + z;\ \zeta \in E\}$. With $0 \leqq \phi \leqq 1/100$ and a segment $J \subset \mathbb{C}$ parallel to the x-axis, we associate the closed segment $J(\phi)$ of the same midpoint as J, parallel to the x-axis and of length $(1 + \phi)\ |J|$. With a positive integer $q \geqq 2$, $0 \leqq \phi \leqq 1/100$ and a segment $J \subset \mathbb{C}$ parallel to the x-axis, we associate

$$J(q,\phi) = \bigcup_{k=1}^{2^{q-1}} \{[J_{2k-1}(\phi) + i\ 2^{-q}|J|]\ \cup J_{2k}(\phi)\}\ ,$$

where $\{J_k\}_{k=1}^{2^q}$ are mutually non-overlapping segments on J of length $2^{-q}|J|$; these are ordered from left to right. The set $J(q,\phi)$ is a union of 2^q closed segments of length $2^{-q}(1 + \phi)\ |J|$. A segment $\Gamma_0 = [0,1]$ is called a crank of type 0. For a finite sequence $\{\phi_j\}_{j=0}^{n}$, $\phi_0 = 0$ of non-negative numbers less than or equal to $1/100$, a finite union Γ of closed segments parallel to the x-axis is called a (fat) crank of type $\{\phi_j\}_{j=0}^{n}$ if there exists a crank $\Gamma' = \bigcup_{k=1}^{\ell} J_k$ ($\{J_k\}_{k=1}^{\ell}$ are components of Γ') of type $\{\phi_j\}_{j=0}^{n-1}$ such that

$$\Gamma = \bigcup_{k=1}^{\ell} J_k(q_k,\ \phi_n)$$

for some ℓ-tuple $(q_1, \ldots, q_\ell)$ of positive integers larger than or equal to 2. We write simply $\Gamma' [_{(q_1, \ldots, q_\ell;\phi_n)} \Gamma$.

Proposition 3.10. Let Γ be a crank of type $\{\phi_j\}_{j=0}^{n}$, $\phi_0 = 0$. Then

$$(3.23) \qquad \gamma(\Gamma) \geqq \text{Const}\ \{ \sum_{\mu=1}^{n} \prod_{j=0}^{\mu} \frac{1}{(1 + \phi_j)} \}^{-1}.$$

At present, the estimate of $\gamma(\Gamma)$ from above is unknown. The method in [29, p. 87] and (3.8) do not yield satisfactory inequalities. The following proof of (3.23) is standard. Let $\{\Gamma_\mu\}_{\mu=0}^{n}$ be $n+1$ cranks such that

$$\Gamma_0 \ [_{(\cdot;\phi_1)}\ \Gamma_1 \ [_{(\cdot;\phi_2)}\ \cdots\ [_{(\cdot;\phi_n)}\ \Gamma_n\ .$$

We put

$$h_\mu(z) = \prod_{j=0}^{\mu} \frac{1}{(1+\phi_j)},$$

$$g_\mu(z) = \operatorname{Im} H_{\Gamma_\mu} h_\mu(z)$$

$$= \frac{1}{\pi} \operatorname{Im} \text{p.v.} \int_{\Gamma_\mu} \frac{h_\mu(\zeta)}{\zeta - z} |d\zeta| \quad (z \in \Gamma_\mu, \ 0 \leqq \mu \leqq n),$$

where Im denotes the imaginary part. Then $g_0 \equiv 0$. We show that, for any $1 \leqq \mu \leqq n$,

$$(3.24) \qquad \|g_\mu\|_{L^\infty(\Gamma_\mu)} \leqq \|g_{\mu-1}\|_{L^\infty(\Gamma_{\mu-1})} + \text{Const} \prod_{j=0}^{\mu} \frac{1}{(1+\phi_j)}.$$

We can write $\Gamma_{\mu-1} = \cup_{k=1}^{\ell} J_k$ with components $\{J_k\}_{k=1}^{\ell}$ of $\Gamma_{\mu-1}$, and can write $\Gamma_\mu = \cup_{k=1}^{\ell} J_k(q_k, \phi_\mu)$ with some ℓ-tuple $(q_1, \dots, q_\ell)$. Let $z_0 \in J_{k_0}(q_{k_0}, \phi_\mu)$ and let z_0^* be the point on J_{k_0} nearest to z_0. Then

$$|g_\mu(z_0) - g_{\mu-1}(z_0^*)| \leqq \Big| \frac{1}{\pi} \operatorname{Im} \int_{J_{k_0}(q_{k_0}, \phi_\mu)} \frac{h_\mu(\zeta)}{\zeta - z_0} |d\zeta|$$

$$- \frac{1}{\pi} \operatorname{Im} \int_{J_{k_0}} \frac{h_{\mu-1}(\zeta)}{\zeta - z_0^*} |d\zeta| \Big|$$

$$+ \Big| \frac{1}{\pi} \sum_{k \neq k_0} \{ \operatorname{Im} \int_{J_k(q_k, \phi_\mu)} \frac{h_\mu(\zeta)}{\zeta - z_0} |d\zeta| - \operatorname{Im} \int_{J_k} \frac{h_{\mu-1}(\zeta)}{\zeta - z_0^*} |d\zeta| \} \Big|$$

$$= L^{(1)} + L^{(2)}, \text{ say.}$$

We can write

$$J_{k_0}(q_{k_0}, \phi_\mu) = \bigcup_{j=1}^{\sigma_0} \gamma_j \qquad (\sigma_0 = 2^{q_{k_0}})$$

with its components $\{\gamma_j\}_{j=1}^{\sigma_0}$; these components are ordered so that the x-coordinate of their midpoints are increasing. Without loss of generality, we may assume that $z_0 \in \cup_{j=1}^{\sigma_0/2} \gamma_{2j-1}$. Since $\{\gamma_{2j}\}_{j=1}^{\sigma_0/2}$ are disjoint and

$$\operatorname{Im} \int_{J_{k_0}} \frac{h_{\mu-1}(\zeta)}{\zeta - z_0^*} |d\zeta| = \operatorname{Im} \int_{\cup_{j=1}^{\sigma_0/2} \gamma_{2j-1}} \frac{h_\mu(\zeta)}{\zeta - z_0} |d\zeta| = 0,$$

we have

$$L^{(1)} \leq \frac{1}{\pi} \left| \operatorname{Im} \int_{\cup_{j=1}^{\sigma_0/2} \gamma_{2j}} \frac{h_\mu(\zeta)}{\zeta - z_0} |d\zeta| \right|$$

$$\leq \frac{1}{\pi} \int_{-\infty}^{\infty} \frac{\sigma_0^{-1} |J_{k_0}|}{(x - \operatorname{Re} z_0)^2 + (\sigma_0^{-1} |J_{k_0}|)^2} dx \, \|h_\mu\|_{L^\infty(\Gamma_\mu)}$$

$$= \prod_{j=0}^{\mu} \frac{1}{(1 + \phi_j)} .$$

For $1 \leq k \leq \ell$, $0 \leq \nu \leq \mu - 1$, $\gamma_k(\nu)$ denotes the component of Γ_ν generating J_k. In particular, $\gamma_k(\mu-1) = J_k$ $(1 \leq k \leq \ell)$. We put

$$L_\nu^{(2)} = \sum_{k \in G_\nu} \left| \int_{J_k(q_k, \phi_\mu)} \frac{h_\mu(\zeta)}{\zeta - z_0} |d\zeta| - \int_{J_k} \frac{h_{\mu-1}(\zeta)}{\zeta - z_0^*} |d\zeta| \right| \quad (1 \leq \nu \leq \mu-1),$$

where

$$G_\nu = \{1 \leq k \leq \ell;\ k \neq k_0,\ \gamma_k(\nu-1) = \gamma_{k_0}(\nu-1),\ \gamma_k(\nu) \neq \gamma_{k_0}(\nu)\}.$$

Then

$$L^{(2)} \leq \sum_{k \neq k_0} \left| \int_{J_k(q_k, \phi_\mu)} \frac{h_\mu(\zeta)}{\zeta - z_0} |d\zeta| - \int_{J_k} \frac{h_{\mu-1}(\zeta)}{\zeta - z_0^*} |d\zeta| \right| = \sum_{\nu=1}^{\mu-1} L_\nu^{(2)} .$$

We now estimate $L_\nu^{(2)}$. Note that $|\gamma_k(\nu)| = |\gamma_{k_0}(\nu)|$ $(k \in G_\nu)$. A geometric observation shows that, for any $k \in G_\nu$,

$$(3.25) \quad \operatorname{dis}(J_k \cup J_k(q_k, \phi_\mu),\ J_{k_0} \cup J_{k_0}(q_{k_0}, \phi_\mu)) \geq \operatorname{dis}(\gamma_k(\nu), \gamma_{k_0}(\nu))$$

$$- 2 |\gamma_{k_0}(\nu)| \ \{(1 + \phi_{\nu+1}) 2^{-2} + (1 + \phi_{\nu+1})(1 + \phi_{\nu+2}) 2^{-4} + \ldots + \prod_{j=\nu+1}^{\mu} (1 + \phi_j)\ 2^{-2(\mu-\nu)}\}$$

$$\geq \operatorname{dis}(\gamma_k(\nu), \gamma_{k_0}(\nu)) - 2 |\gamma_{k_0}(\nu)| \ \{ \frac{1.01}{4} + (\frac{1.01}{4})^2 + \ldots \}$$

$$\geq \operatorname{dis}(\gamma_k(\nu), \gamma_{k_0}(\nu)) - \frac{3}{4} |\gamma_{k_0}(\nu)| .$$

Since

$$\int_{J_k(q_k,\phi_\mu)} h_\mu(\zeta)\ |d\zeta| = \int_{J_k} h_{\mu-1}(\zeta)\ |d\zeta| ,$$

we have, with z_k = (the midpoint of J_k), $Q_k = J_k \cup J_k(q_k,\phi_\mu)$,

$$L_\nu^{(2)} = \sum_{k \in G_\nu} \Big| \int_{J_k(q_k,\phi_\mu)} \{\frac{1}{\zeta - z_0} - \frac{1}{z_k - z_0^*}\} h_\mu(\zeta)\ |d\zeta|$$

$$+ \int_{J_k} \{\frac{1}{z_k - z_0^*} - \frac{1}{\zeta - z_0}\} h_{\mu-1}(\zeta)\ |d\zeta| \Big|$$

$$\leq \text{Const} \sum_{k \in G_\nu} \{(|J_k| + |J_{k_0}|)\ \text{dis}(Q_k, Q_{k_0})^{-2} \int_{J_k} h_{\mu-1}(\zeta)\ |d\zeta|\}$$

$$= \text{Const} \prod_{j=0}^{\mu-1} \frac{1}{(1 + \phi_j)} \sum_{k \in G_\nu} \{(|J_k| + |J_{k_0}|)|J_k|\ \text{dis}(Q_k,Q_{k_0})^{-2}\}.$$

The segment $\gamma_{k_0}(\nu - 1)$ generates $2^{q'_\nu}$ components $\{\lambda_m\}_{m=1}^{2^{q'_\nu}}$ of Γ_ν of length $|\gamma_{k_0}(\nu)|$, where $q'_\nu = \log\{(1 + \phi_\nu)\ |\gamma_{k_0}(\nu-1)|/\ |\gamma_{k_0}(\nu)|\}/\log 2$. We may assume that $\lambda_1 = \gamma_{k_0}(\nu)$. Let

$$G_{\nu,m} = \{k \in G_\nu;\ \ \lambda_m = \gamma_k(\nu)\} \qquad (2 \leq m \leq 2^{q'_\nu}).$$

Then $G_\nu = \cup_{m=2}^{2^{q'_\nu}} G_{\nu,m}$. We have

$$\sum_{k \in G_{\nu,m}} (|J_k| + |J_{k_0}|)|J_k|$$

$$\leq |\lambda_1|\ 2^{-2(\mu-1-\nu)} \prod_{\nu < j \leq \mu-1} (1 + \phi_j) \sum_{k \in G_{\nu,m}} |J_k|$$

$$= |\lambda_1|^2\ 2^{-2(\mu-1-\nu)} \prod_{\nu < j \leq \mu-1} (1 + \phi_j)^2 \leq |\lambda_1|^2\ 2^{-(\mu-1-\nu)},$$

where $\prod_{\nu < j \leq \mu-1} (1 + \phi_j)$ denotes 1 if $\nu = \mu - 1$. Hence a geometric observation and (3.25) show that

$$L_\nu^{(2)} \leqq \text{Const} \prod_{j=0}^{\mu-1} \frac{1}{(1+\phi_j)} \sum_{m=2}^{2^{q'_\nu}} \sum_{k \in G_{\nu,m}} (|J_k| + |J_{k_0}|)|J_k| \, \text{dis}(Q_k, Q_{k_0})^{-2}$$

$$\leqq \text{Const} \prod_{j=0}^{\mu-1} \frac{1}{(1+\phi_j)} \sum_{m=2}^{2^{q'_\nu}} \text{dis}(\lambda_m, \lambda_1)^{-2} \sum_{k \in G_{\nu,m}} (|J_k| + |J_{k_0}|) \, |J_k|$$

$$\leqq \text{Const} \prod_{j=0}^{\mu-1} \frac{1}{(1+\phi_j)} |\lambda_1|^2 \, 2^{-(\mu-1-\nu)} \sum_{m=2}^{2^{q'_\nu}} \text{dis}(\lambda_m, \lambda_1)^{-2}$$

$$\leqq \text{Const} \prod_{j=0}^{\mu-1} \frac{1}{(1+\phi_j)} |\lambda_1|^2 \, 2^{-(\mu-1-\nu)} \sum_{k=1}^{\infty} \left(\frac{1}{|\lambda_1| k} \right)^2$$

$$\leqq \text{Const} \prod_{j=0}^{\mu-1} \frac{1}{(1+\phi_j)} \, 2^{-(\mu-1-\nu)} .$$

Thus

$$|g_\mu(z_0)| \leqq |g_{\mu-1}(z_0^*)| + L^{(1)} + \sum_{\nu=1}^{\mu-1} L_\nu^{(2)}$$

$$\leqq \|g_{\mu-1}\|_{L^\infty(\Gamma_{\mu-1})} + \prod_{j=0}^{\mu} \frac{1}{(1+\phi_j)} + \text{Const} \prod_{j=0}^{\mu-1} \frac{1}{(1+\phi_j)}$$

$$\leqq \|g_{\mu-1}\|_{L^\infty(\Gamma_{\mu-1})} + \text{Const} \prod_{j=0}^{\mu} \frac{1}{(1+\phi_j)} .$$

Since $z_0 \in \Gamma_\mu$ is arbitrary, we obtain (3.24).

By (3.24) and $g_0 \equiv 0$, we have

$$\|\text{Im } H_{\Gamma_n} h_n\|_{L^\infty(\Gamma_n)} \leqq \text{Const} \sum_{\mu=1}^{n} \prod_{j=0}^{\mu} \frac{1}{(1+\phi_j)}$$

$$(= \text{Const } \xi_n, \text{ say}).$$

Evidently,

$$\int_{\Gamma_n} h_n(\zeta) \, |d\zeta| = 1, \quad \|h_n\|_{L^\infty(\Gamma_n)} \leqq 1.$$

Hence we can define a non-negative function h_n^* on Γ_n so that

$$\int_{\Gamma_n} h_n^*(\zeta)\,|d\zeta| = \delta_0/\xi_n ,$$

$$\|h_n^*\|_{L^\infty(\Gamma_n)} + \|\mathrm{Im}\, H_{\Gamma_n} h_n^*\|_{L^\infty(\Gamma_n)} \leq 1,$$

$h_n^*(\zeta) = 0$ at endpoints of each component of Γ_n,

$h_n^*(\zeta)$ is continuously differentiable along Γ_n,

where δ_0 is an absolute constant. Let

$$\hat{h}_n^*(z) = \frac{1}{2\pi i} \int_{\Gamma_n} \frac{h_n^*(\zeta)}{\zeta - z}\,|d\zeta|,$$

$$u_n(z) = \mathrm{Im}\ \hat{h}_n^*(z), \quad v_n(z) = \mathrm{Re}\ \hat{h}_n^*(z),$$

$$f_n(z) = \{1 - \exp(\hat{h}_n^*(z))\} / \{1 + \exp(\hat{h}_n^*(z))\} \qquad (z \notin \Gamma_n). \tag{3.26}$$

Then f_n is analytic outside Γ_n and

$$|f_n'(\infty)| = \frac{1}{2\pi} \int_{\Gamma_n} h_n^*(\zeta)\,|d\zeta| = \delta_0/(2\,\xi_n) .$$

The nontangential limit of $|u_n(z)|$ at each point on Γ_n is dominated by

$$\|\mathrm{Im}\, H_{\Gamma_n} h_n^*\|_{L^\infty(\Gamma_n)} \leq 1.$$

Since $|u_n(z)|$ is subharmonic in Γ_n^c and continuous on $\mathbb{C} \cup \{\infty\}$, we have $\sup_{z \in \Gamma_n^c} |u_n(z)| \leq 1$. Hence, for any $z \notin \Gamma_n$,

$$|f_n(z)|^2 = \frac{1 + \exp(2\,v_n(z)) - 2\exp(v_n(z))\cos(u_n(z))}{1 + \exp(2\,v_n(z)) + 2\exp(v_n(z))\cos(u_n(z))} \leq 1,$$

which shows that $\|f_n\|_{H^\infty} \leq 1$. Consequently,

$$\gamma(\Gamma_n) \geq |f_n'(\infty)| \geq \delta_0/(2\,\xi_n) = \mathrm{Const}\{\sum_{\mu=1}^{n} \prod_{j=0}^{\mu} \frac{1}{(1+\phi_j)}\}^{-1}.$$

This completes the proof of Proposition 3.10.

§3.6. Analytic capacity and integralgeometric quantities

Let $L(r,\theta)$ $(r > 0,\ -\pi < \theta \leqq \pi)$ denote the straight line defined by the equation $x \cos\theta + y \sin\theta = r$. For a compact set $E \subset \mathbb{C}$, we write $N_E(r,\theta) = \#\{E \cap L(r,\theta)\}$, where $\#\{E \cap L(r,\theta)\}$ is the (cardinal) number of elements of $E \cap L(r,\theta)$. For $\varepsilon > 0$ and $0 < \alpha \leqq 1$, we put

$$Cr_\alpha^{(\varepsilon)}(E) = \inf \int_{-\pi}^{\pi} \{\int_0^\infty N_{\partial\{\bigcup_{k=1}^n D(z_k,r_k)\}}(r,\theta)^\alpha\, dr\}\, d\theta,$$

where the infimum is taken over all finite coverings $\{D(z_k,r_k)\}_{k=1}^n$ of E with radii less than ε. We put

$$Cr_\alpha(E) = \lim_{\varepsilon\to 0} Cr_\alpha^{(\varepsilon)}(E) \qquad (0 < \alpha \leqq 1),$$

$$Bu(E) = \lim_{\varepsilon\to 0} \lim_{\alpha\to 0} Cr_\alpha^{(\varepsilon)}(E).$$

If $E \subset D(0,1)$, then $Bu(E)/2\pi$ is called the Buffon needle probability; this is the probability (measured by $dr\, d\theta/2\pi$) of needles $L(r,\theta)$ $(0 < r < 1,\ |\theta| \leqq \pi)$ intersecting with E. Suppose that E is a locally chord-arc compact curve. Then Crofton's formula [49, p.13] shows that $Cr_1(E) = \mathrm{Const}\,|E|$. By (3.7), we have $\gamma(E) \leqq \mathrm{Const}\, Cr_1(E)$. From this point of view, it is interesting to compare $\gamma(\cdot)$ with $Cr_\alpha(\cdot)$ (and with $Bu(\cdot)$). It is known that there exists a compact set E such that $\gamma(E) = 1$ and $Bu(E) = 0$ (cf. Jones-Murai [34]). We shall show

<u>Theorem F.</u> For any $0 < \alpha < 1/2$, there exists a compact set E such that $\gamma(E) = 1$ and $Cr_\alpha(E) = 0$.

Acknowledgement. The author expresses his thanks to Professor Kakutani who communicated (3.28), and Professors Coifman, Steger who suggested to use the Galton-Watson process [30] for the estimate of $Cr_\alpha(\cdot)$. According to Professor Kakutani, various integralgeometric formulae are used for surgeries, since X-rays react to, for example, cancers outside bodies. RSL is the first stopping time of the sun's rays (or needles). Hence it is interesting to try to give an integralgeometric proof of Theorem C.

In this section, we study $Cr_\alpha(\cdot)$. For $0 < \beta < 1$, let $\{X_n^\beta\}_{n=1}^\infty$ be a sequence of independent random variables on the standard probability space $(\Gamma_0, \mathcal{B}, \mathrm{Prob})$ $(\Gamma_0 = [0,1])$ such that

$$\mathrm{Prob}(X_n^\beta = 1) = \mathrm{Prob}(X_n^\beta = -1) = /2,$$

$$\mathrm{Prob}(X_n^\beta = 0) = 1 - \beta \quad (n \geqq 1).$$

We put

$$S_0^\beta = 0,\quad S_n^\beta = X_1^\beta + \ldots + X_n^\beta \qquad (n \geqq 1).$$

This is a model of random walks. We define a Galton-Watson process $\{y_n^\beta\}_{n=0}^\infty$ by

$$(3.27)\qquad y_0^\beta(x) = 1,\quad y_n^\beta(x) = y_{n-1}^\beta(x) + S^\beta_{y_{n-1}^\beta(x)}(x) \quad (n \geqq 1,\ x \in \Gamma_0).$$

Then $\mathrm{Prob}(y_n^\beta \geqq 0) = 1$ $(n \geqq 0)$. We put

$$c_\alpha(n) = 2^{\alpha+1} \int_0^{\pi/2} \{\sum_{k=1}^\infty k^\alpha\, b_k^{(n)}(t(\theta))\}\cos\theta\, d\theta \quad (n \geqq 1,\ 0 \leqq \alpha \leqq 1),$$

where

$$b_k^{(n)}(\beta) = \mathrm{Prob}(y_n^\beta = k),$$

$$t(\theta) = \begin{cases} \tan\theta & (0 \leqq \theta < \arctan 1) \\ |\tan\theta - 2j| & (\arctan(2j-1) \leqq \theta < \arctan(2j+1),\ j \geqq 1). \end{cases}$$

First we show

Lemma 3.11. $\quad c_\alpha(n) \leqq \mathrm{Const}/(\alpha\, n^{1-\alpha}) \quad (n \geqq 1,\ 0 < \alpha \leqq 1).$

The proof of this lemma is standard. The generating function of $\{b_k^{(j)}(\beta)\}_{k=0}^\infty$ is defined by $P_j^\beta(x) = \Sigma_{k=0}^\infty b_k^{(j)}(\beta)\, x^k$. Then

$$(3.28)\qquad P_j^\beta(x) = P_{j-1}^\beta\left(\frac{\beta}{2} + (1-\beta)x + \frac{\beta}{2}x^2\right).$$

In effect, (3.27) shows that, for any $k \geqq 0$,

$$(3.29)\qquad b_k^{(j)}(\beta) = \mathrm{Prob}(y_j^\beta = k) = \sum_{\ell=0}^\infty \mathrm{Prob}(y_{j-1}^\beta = \ell)\,\mathrm{Prob}(S_\ell^\beta = k - \ell)$$

$$= \sum_{\ell=0}^\infty b_\ell^{(j-1)}(\beta)\ \Sigma_{(\ell)} \frac{\ell!}{\ell_1!\,\ell_2!\,\ell_3!} \left(\frac{\beta}{2}\right)^{\ell_1}(1-\beta)^{\ell_2}\left(\frac{\beta}{2}\right)^{\ell_3},$$

where $\Sigma_{(\ell)}$ is the summation taken over all triples (ℓ_1, ℓ_2, ℓ_3) of non-negative integers such that $\ell_1 + \ell_2 + \ell_3 = \ell$, $\ell_2 + 2\,\ell_3 = k$. The x^k-coefficient of $P_j^\beta(x)$ is $b_k^{(j)}(x)$ and the x^k-coefficient of $P_{j-1}^\beta(\frac{\beta}{2} + (1-\beta)x + \frac{\beta}{2}x^2)$ is equal to the last quantity in (3.29). Thus (3.28) holds. Let

$$v_j(\beta) = \int_{\phi_j(\beta)}^{1} (1 - x)^{-\alpha} \frac{\partial}{\partial x} P^{\beta}_{n-j}(x)\, dx \quad (0 \leqq j \leqq n),$$

where $\{\phi_j(\beta)\}_{j=0}^{n}$ are inductively defined by $\phi_0(\beta) = 0$,

$$\phi_j(\beta) = \frac{\beta}{2} + (1 - \beta)\, \phi_{j-1}(x) + \frac{\beta}{2}\, \phi_{j-1}(x)^2 .$$

Since

$$\int_0^1 (1 - x)^{-\alpha}(k\, x^{k-1})\, dx \geqq \int_{1-(1/k)}^{1} (1 - x)^{-\alpha}(k\, x^{k-1})\, dx$$

$$\geqq \quad \text{Const}\ \frac{k^{\alpha}}{1 - \alpha} \qquad (k \geqq 2),$$

we have

$$\sum_{k=1}^{\infty} k^{\alpha}\, b_k^{(n)}(\beta) \leqq \text{Const}\ (1 - \alpha)\ v_0(\beta).$$

Since

$$(1 - x)^{-\alpha} \leqq \{1 - (\frac{\beta}{2} + (1 - \beta)x + \frac{\beta}{2} x^2)\}^{-\alpha} \quad (0 \leqq x \leqq 1),$$

(3.28) shows that

$$v_j(\beta) = \int_{\phi_j(\beta)}^{1} (1 - x)^{-\alpha}\{(1-\beta) + \beta x\} \frac{\partial}{\partial x} P^{\beta}_{n-j-1}(\frac{\beta}{2} + (1-\beta)x + \frac{\beta}{2} x^2) dx$$

$$\leqq \int_{\phi_j(\beta)}^{1} \{1-(\frac{\beta}{2} + (1-\beta)x + \frac{\beta}{2} x^2)\}^{-\alpha}\{(1-\beta) + \beta x\} \frac{\partial}{\partial x} P^{\beta}_{n-j-1}(\frac{\beta}{2} + (1-\beta)x + \frac{\beta}{2} x^2) dx$$

$$= \quad v_{j+1}(\beta),$$

and hence

$$v_0(\beta) \leqq v_1(\beta) \leqq \ldots \leqq v_n(\beta)$$

$$= \int_{\phi_n(\beta)}^{1} (1 - x)^{-\alpha} \frac{\partial}{\partial x} P_0^{\beta}(x) dx = \frac{1}{1 - \alpha} (1 - \phi_n(\beta))^{1-\alpha}.$$

We have easily $|1 - \phi_n(\beta)| \leqq \text{Const}/(\beta n)$. Thus

$$\sum_{k=1}^{\infty} k^{\alpha}\, b_k^{(n)}(\beta) \leqq \text{Const}/(\beta n)^{1-\alpha} .$$

Consequently, we have, with $\theta_j = \arctan j \ (j \geqq 0)$,

$$c_\alpha(n) \leq \text{Const } n^{\alpha-1} \int_0^{\pi/2} t(\theta)^{\alpha-1} \cos\theta \, d\theta$$

$$= \text{Const } n^{\alpha-1}\{ \int_0^{\theta_1} (\tan\theta)^{\alpha-1} \cos\theta \, d\theta + \sum_{j=1}^{\infty} \int_{\theta_{2j-1}}^{\theta_{2j+1}} |\tan\theta - 2j|^{\alpha-1} \cos\theta \, d\theta\}$$

$$= \text{Const } n^{\alpha-1}\{ \int_0^1 \frac{y^{\alpha-1}}{(1+y^2)^{3/2}} dy$$

$$+ \sum_{j=1}^{\infty} \int_{-1}^{1} \frac{y^{\alpha-1}}{\{1+(y+2j)^2\}^{3/2}} dy \leq \text{Const}/(\alpha\, n^{1-\alpha}).$$

This completes the proof of Lemma 3.11.

Let Γ' and Γ be two cranks such that n = (the degree of Γ) - (the degree of Γ') $\geqq 1$. If there exist an n-tuple $(\Lambda_1, \ldots, \Lambda_n)$ of cranks such that

$$\Gamma' = \Lambda_1 \underset{(\mathcal{Q}_1;\ \cdot)}{[} \Lambda_2 \underset{(\mathcal{Q}_2;\ \cdot)}{[} \cdots \underset{(\mathcal{Q}_n;\ \cdot)}{[} \Lambda_n = \Gamma$$

for some n-tuple $(\mathcal{Q}_1, \ldots, \mathcal{Q}_n)$, $\mathcal{Q}_\mu = \{q_k^{(\mu)}\}_{k=1}^{\ell_\mu}$ of finite sequences of positive integers, we write $\Gamma' [[\Gamma$ and

$$\iota(\Gamma', \Gamma) = \min \{q_k^{(\mu)};\ 1 \leqq k \leqq \ell_\mu,\ 1 \leqq \mu \leqq n\}.$$

Note that, for any crank Γ,

$$Cr_\alpha(\Gamma) = 2^\alpha \int_{-\pi}^{\pi} \{\int_0^\infty N_\Gamma(r,\theta)^\alpha \, dr\} \, d\theta \qquad (0 \leqq \alpha \leqq 1).$$

Lemma 3.12. For $0 \leqq \varepsilon \leqq 1$, $0 \leqq \alpha \leqq 1$, $\ell_0 \geqq 1$ and $n \geqq 1$, there exists a crank Γ_n^* of degree n such that

$$(3.30) \qquad \iota(\Gamma_0, \Gamma_n^*) \geqq \ell_0, \quad |Cr_\alpha(\Gamma_n^*) - c_\alpha(n)| \leqq \varepsilon,$$

where $\Gamma_0 = [0,1]$.

Proof. For a finite increasing sequence $\mathcal{Q} = \{q_k\}_{k=1}^{\ell}$ of positive integers, we put $\text{gap}(\mathcal{Q}) = \min \{q_k - q_{k-1};\ 1 \leqq k \leqq \ell\}$, where $q_0 = 0$. Let $\mathbb{Q}_n = (\mathcal{Q}_1, \ldots, \mathcal{Q}_n)$, $\mathcal{Q}_\mu = \{q_k^{(\mu)}\}_{k=1}^{\ell_\mu}$ be an n-tuple of finite increasing sequences of positive integers such that

$$\ell_1 = 1, \quad \ell_\mu = 2^{q_1^{(\mu)}} + 2^{q_2^{(\mu)}} + \ldots + 2^{q_{\ell_{\mu-1}}^{(\mu)}} \qquad (2 \leqq \mu \leqq n),$$

we put $\text{gap}(\mathbb{Q}_n) = \min\{\text{gap}(\mathcal{Q}_\mu);\ 1 \leqq \mu \leqq n\}$. With $\mathbb{Q}_n$, we associate n cranks

$\Gamma_\mu^* = \Gamma_\mu^*(\mathcal{Q}_1, \dots, \mathcal{Q}_\mu)$ $(1 \leqq \mu \leqq n)$ as follows. Let $\Gamma_1^* = \Gamma_1^*(\mathcal{Q}_1) = \Gamma_0(q_1^{(1)},0)$ ($\Gamma_0 = [0,1]$). Suppose that $\Gamma_1^*, \Gamma_2^*, \dots, \Gamma_{\mu-1}^*$ have been defined so that Γ_k^* has ℓ_k components $(1 \leqq k \leqq \mu-1)$. Then $\Gamma_{\mu-1}^*$ is expressed in the form $\Gamma_{\mu-1}^* = \cup_{k=1}^{\ell_{\mu-1}} J_k^{(\mu-1)}$ with its components $\{J_k^{(\mu-1)}\}_{k=1}^{\ell_{\mu-1}}$; these are ordered so that the x-coordinates of their midpoints are increasing. We put

$$\Gamma_\mu^* = \Gamma_\mu^*(\mathcal{Q}_1, \dots, \mathcal{Q}_\mu) = \bigcup_{k=1}^{\ell_{\mu-1}} J_k^{(\mu-1)}(q_k^{(\mu-1)},0).$$

The set $\Gamma_n^*(\mathbb{Q}_n)$ is a crank of degree n. We now study $Cr_\alpha(\Gamma_n^*(\mathbb{Q}_n))$. We have, with $\theta_j = \arctan j$, $0 \leqq \theta_j < \pi/2$ $(j \geqq 0)$,

$$2^{-\alpha} Cr_\alpha(\Gamma_n^*(\mathbb{Q}_n)) = \int_{-\pi}^{\pi}\{\int_0^\infty N_{\Gamma_n^*(\mathbb{Q}_n)}(r,\theta)^\alpha \, dr\} \, d\theta$$

$$= \sum_{j=1}^{\infty} \{\int_{\theta_{j-1}}^{\theta_j} \int_0^\infty + \int_{-\theta_j}^{-\theta_{j-1}} \int_0^\infty\} + \int_{\pi/2 < |\theta| < \pi} \int_0^\infty$$

$$(= \sum_{j=1}^{\infty} \{d_j(\mathbb{Q}_n) + d_{-j}(\mathbb{Q}_n)\} + d_0(\mathbb{Q}_n), \text{ say}).$$

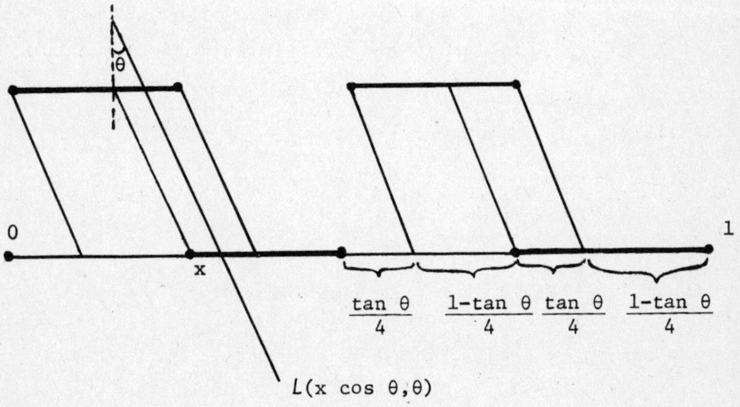

For $0 < \theta < \theta_1$, we put $\tilde{b}_1^{(0)}(\theta) = 0$,

$$\tilde{b}_k^{(\mu)}(\theta) = \tilde{b}_k^{(\mu)}(\theta;\mathcal{Q}_1, \dots, \mathcal{Q}_\mu)$$

$$= |x \in \Gamma_0;\ N_{\Gamma_\mu^*(\mathcal{Q}_1, \dots, \mathcal{Q}_\mu)}(x \cos\theta,\theta) = k| \qquad (k \geqq 0,\ 1 \leqq \mu \leqq n).$$

Then $\tilde{b}_k^{(\mu)}(\theta) = 0 \quad (k \geq 2^{\mu} + 1\ ,\ 0 \leq \mu \leq n)$. We have

$$\tilde{b}_1^{(0)}(\theta) = 1 = \mathrm{Prob}(y_0^{\tan\theta} = 1)$$

$$\tilde{b}_k^{(1)}(\theta) = \mathrm{Prob}(y_1^{\tan\theta} = k) \qquad (0 \leq k \leq 2).$$

Let $1 \leq j \leq 2^{\mu-1}$, $2 \leq \mu \leq n$. To a component V of $\{x \in \Gamma_0;\ N_{\Gamma^*_{\mu-1}}(x\cos\theta,\theta) = j\}$, there correspond j components $J_{\nu_1}^{(\mu-1)}, \ldots, J_{\nu_j}^{(\mu-1)}$ of $\Gamma^*_{\mu-1}$ which intersect with $L(x\cos\theta,\theta)$ for all $x \in V$; these are ordered so that the x-coordinates of their midpoints are increasing. If $q_{\nu_1}^{(\mu)}$ is sufficiently large, then

$$\Big|\,|x \in V;\ \#\{J_{\nu_1}^{(\mu-1)}(q_{\nu_1}^{(\mu)},0) \cap L(x\cos\theta,\theta)\} = k\,|$$
$$-\ |V|\ \mathrm{Prob}(S_1^{\tan\theta} = k-1)\Big|$$

is sufficiently small for all $0 \leq k \leq 2$. If $q_{\nu_1}^{(\mu)},\ q_{\nu_2}^{(\mu)} - q_{\nu_1}^{(\mu)}$ are sufficiently large, then

$$\Big|\,|x \in V;\ \#\{\big(J_{\nu_1}^{(\mu-1)}(q_{\nu_1}^{(\mu)},0) \cup J_{\nu_2}^{(\mu-1)}(q_{\nu_2}^{(\mu)},0)\big) \cap L(x\cos\theta,\theta)\} = k\,\Big|$$
$$-\ |V|\ \mathrm{Prob}(S_2^{\tan\theta} = k-2)\Big|$$

is sufficiently small for all $0 \leq k \leq 4$. Repeating this argument, we see that, if $q_{\nu_1}^{(\mu)},\ q_{\nu_2}^{(\mu)} - q_{\nu_1}^{(\mu)},\ \ldots,\ q_{\nu_j}^{(\mu)} - q_{\nu_{j-1}}^{(\mu)}$ are sufficiently large, then

$$\Big|\,|x \in V;\ \#\{\Gamma^*_{\mu} \cap L(x\cos\theta,\theta)\} = k| - |V|\ \mathrm{Prob}(S_j^{\tan\theta} = k-j)\Big|$$

is sufficiently small for all $0 \leq k \leq 2j$. Hence, if $\mathrm{gap}(\mathcal{Q}_\mu)$ is sufficiently large, then

$$\Big|\,|x \in \Gamma_0;\ N_{\Gamma^*_{\mu-1}}(x\cos\theta,\theta) = j,\ N_{\Gamma^*_{\mu}}(x\cos\theta,\theta) = k|$$
$$-\ \tilde{b}_j^{(\mu-1)}(\theta)\ \mathrm{Prob}(S_j^{\tan\theta} = k-j)\Big|$$

is sufficiently small for all $0 \leq k \leq 2j$. If $\mathrm{gap}(\mathcal{Q}_\mu)$ is sufficiently large, then

$$|\ x \in \Gamma_0;\ N_{\Gamma^*_{\mu-1}}(x\cos\theta,\theta) = 0,\ \ N_{\Gamma^*_{\mu}}(x\cos\theta,\theta) \geq 1\ |$$

$$(= \left| |x \in \Gamma_0;\ N_{\Gamma^*_{\mu-1}}(x\cos\theta,\theta) = 0,\ \ N_{\Gamma^*_{\mu}}(x\cos\theta,\theta) \geqq 1| - \tilde{b}_0^{(\mu-1)}(\theta)\ \mathrm{Prob}(S_0^{\tan\theta} \geqq 1) \right|)$$

is sufficiently small. Thus there exists a positive integer $p_\mu(\theta)$ such that, if $\mathrm{gap}(\mathcal{Q}_\mu) \geqq p_\mu(\theta)$, then

$$\left| |x \in \Gamma_0;\ N_{\Gamma^*_{\mu-1}}(x\cos\theta,\theta) = j,\ \ N_{\Gamma^*_{\mu}}(x\cos\theta,\theta) = k| - \tilde{b}_j^{(\mu-1)}(\theta)\ \mathrm{Prob}(S_j^{\tan\theta} = k-j) \right| \leqq \varepsilon\, 2^{-3n^3}$$

for all $0 \leqq k \leqq 2j$, $0 \leqq j \leqq 2^{\mu-1}$. This yields that

$$\left| \tilde{b}_k^{(\mu)}(\theta) - \sum_{j=0}^{2^{\mu-1}} \tilde{b}_j^{(\mu-1)}(\theta)\ \mathrm{Prob}(S_j^{\tan\theta} = k-j) \right|$$

$$= \left| \sum_{j=0}^{2^{\mu-1}} \{ |x \in \Gamma_0;\ N_{\Gamma^*_{\mu-1}}(x\cos\theta,\theta) = j,\ N_{\Gamma^*_{\mu}}(x\cos\theta,\theta) = k| - \tilde{b}_j^{(\mu-1)}(\theta)\ \mathrm{Prob}(S_j^{\tan\theta} = k-j) \} \right|$$

$$\leqq \varepsilon(2^{\mu-1}+1)2^{-3n^3} \leqq \mathrm{Const}\ \varepsilon\ 2^{-2n^3} \qquad (0 \leqq k \leqq 2^\mu).$$

Put $p(\theta) = \max\{p_\mu(\theta);\ 1 \leqq \mu \leqq n\}$. If $\mathrm{gap}(\mathcal{Q}_n) \geqq p(\theta)$, then

$$\sum_{k=0}^{2^n} |\tilde{b}_k^{(n)}(\theta) - b_k^{(n)}(\tan\theta)| = \sum_{k=0}^{2^n} |\tilde{b}_k^{(n)}(\theta) - \mathrm{Prob}(y_n^{\tan\theta} = k)|$$

$$\leqq \sum_{k=0}^{2^n} \left| \sum_{j=0}^{2^{n-1}} \tilde{b}_j^{(n-1)}(\theta)\ \mathrm{Prob}(S_j^{\tan\theta} = k-j) - \sum_{j=0}^{2^{n-1}} \mathrm{Prob}(y_{n-1}^{\tan\theta} = j)\mathrm{Prob}(S_j^{\tan\theta} = k-j) \right| + \mathrm{Const}\ \varepsilon\ (2^n+1)2^{-2n^3}$$

$$\leqq (2^n+1) \sum_{j=0}^{2^{n-1}} |\tilde{b}_j^{(n-1)}(\theta) - \mathrm{Prob}(y_{n-1}^{\tan\theta} = j)| + \mathrm{Const}\ \varepsilon(2^n+1)2^{-2n^3}$$

$$\leqq \ldots \leqq \prod_{\mu=2}^{n} (2^\mu+1) \sum_{j=0}^{2} |\tilde{b}_j^{(1)}(\theta) - \mathrm{Prob}(y_1^{\tan\theta} = j)|$$

$$+ \mathrm{Const}\ \ \varepsilon\{ \sum_{\mu=2}^{n} (2^\mu+1)\ldots(2^n+1)\}\ 2^{-2n^3} \leqq \mathrm{Const}\ \varepsilon\ 2^{-n^3},$$

which shows that

$$|\int_0^\infty N_{\Gamma_n^*(\mathbb{Q}_n)} (r,\theta)^\alpha \, dr - \sum_{k=1}^{\infty} k^\alpha b_k^{(n)}(\tan\theta)\cos\theta|$$

$$= |\int_0^1 N_{\Gamma_n^*} (x\cos\theta,\theta)^\alpha \cos\theta \, dx - \sum_{k=1}^{2^n} k^\alpha b_k^{(n)}(\tan\theta)\cos\theta|$$

$$= |\sum_{k=1}^{2^n} k^\alpha \{\tilde{b}_k^{(n)}(\theta) - b_k^{(n)}(\tan\theta)\} \cos\theta|$$

$$\leqq \text{Const } \varepsilon \, 2^{\alpha n} \, 2^{-n^3} \leqq \text{Const } \varepsilon .$$

There exists a positive integer p_1 such that the measure of $F = \{0 < \theta < \theta_1;\ p(\theta) > p_1\}$ is less than $\varepsilon\, 2^{-n}$. If $\text{gap}(\mathbb{Q}_n) \geqq p_1$, then

$$|d_1(\mathbb{Q}_n) - \int_0^{\theta_1} \sum_{k=1}^{\infty} k^\alpha b_k^{(n)}(\tan\theta)\cos\theta \, d\theta|$$

$$= |\int_0^{\theta_1} \{\int_0^\infty N_{\Gamma_n^*} (r,\theta)^\alpha \, dr - \sum_{k=1}^{\infty} k^\alpha b_k^{(n)}(\tan\theta)\cos\theta\} \, d\theta|$$

$$= |\{\int_F + \int_{(0,\theta_1)-F}\} \{\quad\} \, d\theta|$$

$$\leqq \text{Const } 2^{\alpha n} |F| + \text{Const } \varepsilon \leqq \text{Const } \varepsilon .$$

For an integer $j \neq 1$, we can choose, in the same manner as above, a positive integer p_j such that, if $\text{gap}(\mathbb{Q}_n) \geqq p_j$, then

$$|d_j(\mathbb{Q}) - \int_{\theta_{|j|-1}}^{\theta_{|j|}} \sum_{k=1}^{\infty} k^\alpha b_k^{(n)}(t(\theta))\cos\theta \, d\theta| \leqq \varepsilon/(1+j^2),$$

where $\theta_{-1} = 0$. This shows that if $\text{gap}(\mathbb{Q}_n)$ is sufficiently large, then (3.30) holds. Q.E.D.

§3.7. Proof of Theorem F ([34])

Here are three lemmas necessary for the proof. From now on, we fix $0 < \alpha < 1/2$.

Lemma 3.13. For $\ell_0 \geqq 1$ and $n \geqq 1$, there exist a crank Γ_n^* of degree n and a non-negative function w_n^* on Γ_n^* such that w_n^* is a constant on each component of Γ_n^*,

$$(3.31)\quad \begin{cases} \iota(\Gamma_0, \Gamma_n^*) \geqq \ell_0, \quad Cr_\alpha(\Gamma_n^*) \leqq C_1/(\alpha\, n^{1-\alpha}), \\ \|w_n^*\|_{L^1(\Gamma_n^*)} = 1, \quad \|w_n^*\|_{L^\infty(\Gamma_n^*)} \leqq C_1, \ \|\mathrm{Im}\, H_{\Gamma_n^*} w_n^*\|_{L^\infty(\Gamma_n^*)} \leqq C_n\sqrt{n}, \end{cases}$$

where C_1 is an absolute constant.

Proof. Lemmas 3.11 and 3.12 show that there exists a crank Γ_n^* of degree n satisfying the first two inequalities in (3.31). Inequality (3.13) shows that

$$\|H_{\Gamma_n^*}\|_{L^2(\Gamma_n), L^2(\Gamma_n)} \leqq \mathrm{Const}\sqrt{n},$$

which yields

$$\|H_{\Gamma_n^*}\|_{L^1(\Gamma_n), L_w^1(\Gamma_n)} \leqq \mathrm{Const}\sqrt{n}.$$

Thus, in the same manner as in the proof of Theorem D, we obtain a non-negative function on Γ_n^* satisfying the last three inequalities in (3.31). Taking the mean over each component of Γ_n^*, we obtain the required function w_n^*. Q.E.D.

Lemma 3.14. Let $\ell_0 \geqq 1$ and $n \geqq 1$. Let Γ_m be a crank of type $\{\phi_j\}_{j=0}^m$ and w_m be a non-negative function on Γ_m such that w_m is a constant on each component of Γ_m. Then there exists a crank Γ_{m+n} of type $\{\phi_j\}_{j=0}^{m+n}$ with $\phi_j = 0$ $(m+1 \leqq j \leqq m+n)$ and a non-negative function w_{m+n} on Γ_{m+n} such that w_{m+n} is a constant on each component of Γ_{m+n},

$$\Gamma_m \,[[\, \Gamma_{m+n}, \ \iota(\Gamma_m, \Gamma_{m+n}) \geqq \ell_0,$$

$$Cr_\alpha(\Gamma_{m+n}) \leqq C_1\, |\Gamma_m|/(\alpha\, n^{1-\alpha}),$$

$$\|w_{m+n}\|_{L^1(\Gamma_{m+n})} = \|w_m\|_{L^1(\Gamma_m)}, \ \|w_{m+n}\|_{L^\infty(\Gamma_{m+n})} \leqq C_1\, \|w_m\|_{L^\infty(\Gamma_m)},$$

$$(3.32)\quad \|\mathrm{Im}\, H_{\Gamma_{m+n}} w_{m+n}\|_{L^\infty(\Gamma_{m+n})} \leqq \|\mathrm{Im}\, H_{\Gamma_m} w_m\|_{L^\infty(\Gamma_m)} + C_2\sqrt{n}\, \|w_m\|_{L^\infty(\Gamma_m)},$$

where C_1 is the constant in Lemma 3.13 and C_2 is an absolute constant.

Proof. We can write $\Gamma_m = \bigcup_{k=1}^{\ell} J_k$ with its components $\{J_k\}_{k=1}^{\ell}$. Let $\{z_k\}_{k=1}^{\ell}$ be the left endpoints of $\{J_k\}_{k=1}^{\ell}$, respectively. We put

$$\Gamma_{m+n} = \bigcup_{k=1}^{\ell} \Lambda_k, \quad \Lambda_k = [|J_k|\Gamma_n^* + z_k],$$

$$w_{m+n}(z) = w_n^*((z-z_k)/|J_k|)\, w_m(z_k) \qquad (z \in \Lambda_k, \ 1 \leqq k \leqq \ell),$$

where Γ_n^* and w_n^* are the crank and the function in Lemma 3.13, respectively. Then Γ_{m+n} is a crank of type $\{\phi_j\}_{j=0}^{m+n}$ such that Γ_m [[Γ_{m+n}, $\iota(\Gamma_m, \Gamma_{m+n}) \geqq \ell_0$. The second inequality in (3.31) shows that

$$Cr_\alpha(\Gamma_{m+n}) \leqq \sum_{k=1}^{\ell} Cr_\alpha(\Lambda_k) = \sum_{k=1}^{\ell} |J_k|\, Cr_\alpha(\Gamma_n^*) \leqq C_1\, |\Gamma_m|/(\alpha\, n^{1-\alpha}).$$

In the same manner as in Proposition 3.10, we have (3.32). Q.E.D.

In the same manner, we have

Lemma 3.15. Let $\ell_0 \geqq 1$, $n \geqq 1$, Γ_m be a crank of type $\{\phi_j\}_{j=0}^{m}$, w_m be a non-negative function on Γ_m such that w_m is a constant on each component of Γ_m, and let $\{\phi_j\}_{j=m+1}^{m+n}$ be non-negative numbers less than or equal to 1/100. Then there exist a crank of type $\{\phi_j\}_{j=0}^{m+n}$ and a non-negative function w_{m+n} on Γ_{m+n} such that w_{m+n} is a constant on each component of Γ_{m+n}, Γ_m [[Γ_{m+n}, $\iota(\Gamma_m, \Gamma_{m+n}) \geqq \ell_0$,

$$\|w_{m+n}\|_{L^1(\Gamma_{m+n})} = \|w_m\|_{L^1(\Gamma_m)},$$

$$\|w_{m+n}\|_{L^\infty(\Gamma_{m+n})} \leqq \|w_m\|_{L^\infty(\Gamma_m)} \Big/ \prod_{\mu=m+1}^{m+n} (1 + \phi_\mu),$$

$$\|\mathrm{Im}\ H_{\Gamma_{m+n}} w_{m+n}\|_{L^\infty(\Gamma_{m+n})} \leqq \|\mathrm{Im}\ H_{\Gamma_m} w_m\|_{L^\infty(\Gamma_m)} + \|w_m\|_{L^\infty(\Gamma_m)} \sum_{j=m+1}^{m+n} \{1/\prod_{\mu=m+1}^{j} (1 + \phi_\mu)\}.$$

We now construct the required compact set E. Choose a positive number β_0 and a positive integer $n_0 \geqq 2$ so that $\beta_0/2 < 1 - (1/n_0)$ and $\beta_0(1 - \alpha) > 1$. Let p_m be the integral part of $(101/100)^{\beta_0 m}$ $(m \geqq 1)$. We define a sequence $\{m_k\}_{k=0}^{\infty}$ of non-negative integers by $m_0 = 0$, $m_1 = n_0$

$$m_{k+1} = n_0 m_k + p_{n_0 m_k} \qquad (k \geqq 1),$$

and define a sequence $\{\phi_j\}_{j=0}^{\infty}$ of non-negative numbers by

$$\phi_j = \begin{cases} 0 & (0 < j \leqq m_1) \\ 1/100 & (m_k < j \leqq n_0 m_k,\ k \geqq 1) \\ 0 & (n_0 n_k < j \leqq m_{k+1},\ k \geqq 1). \end{cases}$$

Let $\mathbb{L} = \{\ell_k\}_{k=1}^{\infty}$ be an increasing sequence of positive integers which will be determined later. Using Lemma 3.14 with $\ell_0 = \ell_1$, $n = m_1$, Γ_0 and $w_0 \equiv 1$, we obtain a crank Γ_{m_1} and a non-negative function w_{m_1}. Using Lemma 3.15 with $\ell_0 = \ell_1$, $n = (n_0 - 1)m_1$, Γ_{m_1}, w_{m_1} and $\{\phi_j\}_{j=m_1+1}^{n_0 m_1}$, we obtain a crank $\Gamma_{n_0 m_1}$ and a non-negative function $w_{n_0 m_1}$. Using Lemma 3.14 with $\ell_0 = \ell_2$, $n = p_{n_0 m_1}$, $\Gamma_{n_0 m_1}$ and $w_{n_0 m_1}$, we obtain a crank Γ_{m_2} and a non-negative function w_{m_2}. Repeating this argument, we obtain a sequence $\{\Gamma_{m_k}\}_{k=1}^{\infty}$ of cranks and a sequence $\{w_{m_k}\}_{k=1}^{\infty}$ of non-negative functions such that, for $k \geqq 2$,

$$\Gamma'_k\ [[\ \Gamma'_{k+1},\ \iota(\Gamma'_k, \Gamma'_{k+1}) \geqq \ell_k,$$

$$\|w_{m_k}\|_{L^1(\Gamma'_k)} = 1, \quad \|w_{m_k}\|_{L^\infty(\Gamma'_k)} \leqq C_0^k \Big/ \prod_{\mu=0}^{n_0 m_{k-1}} (1 + \phi_\mu),$$

$$\|\mathrm{Im}\, H_{\Gamma'_k} w_{m_k}\|_{L^\infty(\Gamma'_k)} \leqq C_0 \sqrt{m_1}$$

$$+ \sum_{\nu=1}^{k-1} C_0^{\nu} \sum_{j=m_\nu+1}^{n_0 m_\nu} \{1/\prod_{\mu=0}^{j} (1 + \phi_\mu)\} + \sum_{\nu=1}^{k-1} C_0^{\nu+1} \sqrt{p_{n_0 m_\nu}} \Big/ \prod_{\mu=1}^{n_0 m_\nu} (1 + \phi_\mu),$$

$$\mathrm{Cr}_\alpha(\Gamma'_k) \leqq C_0 \prod_{\mu=0}^{n_0 m_{k-1}} (1 + \phi_\mu)/(\alpha\, p_{n_0 m_{k-1}}^{1-\alpha}),$$

where $\Gamma'_k = \Gamma_{m_k}$ and $C_0 = \max\{C_1, C_2\}$. We put $E(\mathbb{L}) = \bigcap_{j=1}^{\infty} \overline{\bigcup_{k=j}^{\infty} \Gamma'_k}$, and show that $\gamma(E(\mathbb{L})) \geqq \mathrm{Const.}$ Let $k \geqq 2$. Then $\|w_{m_k}\|_{L^1(\Gamma'_k)} = 1$. Since $m_\nu \geqq n_0 m_{\nu-1}$ $(\nu \geqq 2)$ and $m_1 = n_0$, we have $m_\nu \geqq n_0^{\nu}$ $(\nu \geqq 1)$, and hence

$$\|w_{m_k}\|_{L^\infty(\Gamma'_k)} \leqq C_0^k \Big/ \prod_{\mu=m_{k-1}+1}^{n_0 m_{k-1}} (1 + \phi_\mu)$$

$$= C_0^k \left(\frac{101}{100}\right)^{-(n_0-1)m_{k-1}} \leqq \mathrm{Const.}$$

Since

$$\sqrt{p_{n_0 m_\nu}} \Big/ \prod_{\mu=0}^{n_0 m_\nu} (1+\phi_\mu) \leqq \text{Const}\left(\frac{101}{100}\right)^{\beta_0 n_0 m_\nu/2} \left(\frac{101}{100}\right)^{-(n_0-1)m_\nu}$$

$$= \text{Const}\left(\frac{101}{100}\right)^{-\{(1-(1/n_0))-(\beta_0/2)\}n_0 m_\nu} \qquad (\nu \geqq 1),$$

we have

$$\|\text{Im}\, H_{\Gamma_k'}\, w_{m_k}\|_{L^\infty(\Gamma_k')} \leqq \text{Const.}$$

Thus, in the same manner as in (3.26), we obtain an analytic function $f_k \in H^\infty(\Gamma_k'^{c})$ such that

$$\|f_k\|_{H^\infty} \leqq 1, \quad |f_k'(\infty)| \geqq \text{Const.}$$

Since $k \geqq 2$ is arbitrary, we obtain an analytic function $f \in H^\infty(E(\mathbb{L})^c)$ such that

$$\|f\|_{H^\infty} \leqq 1, \quad |f'(\infty)| \geqq \text{Const},$$

which shows that $\gamma(E(\mathbb{L})) \geqq \text{Const}$.

Let

$$\varepsilon_k = (C_0/\alpha)\left(\frac{101}{100}\right)^{(1-(1-\alpha)\beta_0)n_0 m_k} \qquad (k \geqq 2).$$

Then $\lim_{k\to\infty} \varepsilon_k = 0$ and

$$\text{Cr}_\alpha(\Gamma_k') \leqq (2C_0/\alpha)\left(\frac{101}{100}\right)^{n_0 m_{k-1}} \left(\frac{101}{100}\right)^{-(1-\alpha)\beta_0 n_0 m_{k-1}}$$

$$\leqq 2\,\varepsilon_{k-1}\,.$$

We can inductively choose $\mathbb{L}^0 = \{\ell_k^0\}_{k=1}^\infty$ so that, for any $k \geqq 2$, $\text{Cr}_\alpha^{(1/k)}(E(\mathbb{L}^0)) \leqq 2\,\varepsilon_{k-1}$, which shows that $\text{Cr}_\alpha(E(\mathbb{L}^0)) = 0$. Thus $E = E(\mathbb{L}^0)$ satisfies $\gamma(E) > 0$ and $\text{Cr}_\alpha(E) = 0$.

Remark 3.16. Throughout the note, we use Theorem D to estimate $\gamma(\cdot)$ from below. Here is a weaker inequality than Theorem D. Let Γ be a locally chord-arc curve. Then

$$\gamma(\Gamma) \geqq \text{Const}\, \left|\int_\Gamma f\, dz\right|^2 \Big/ \{\|f\|^2_{L^2(\Gamma)} + \|H_\Gamma(f\, dz/|dz|)\|^2_{L^2(\Gamma)}\}$$

(cf. [29, p. 19]). This is also useful to estimate $\gamma(\Gamma)$ from below. In effect, we can deduce (3.23) and $\{\|C[a]\|_{2,2};\ a \in L^\infty_{\text{real}}\} = \infty$ from this inequality.

APPENDIX I. AN EXTREMAL PROBLEM

For $s_1, \ldots, s_n \in \mathbb{R}$, we define

$$T_{s_1, \ldots, s_n}(x,y) = 1/\{(x-y) + i(A_{s_1, \ldots, s_n}(x) - A_{s_1, \ldots, s_n}(y)\},$$

where

$$A_{s_1, \ldots, s_n}(x) = \begin{cases} 0 & x \notin I_0 = [0,1) \\ s_k & (\frac{k-1}{n} \leqq x < \frac{k}{n}, \quad 1 \leqq k \leqq n). \end{cases}$$

Put

$$(4.1) \qquad ex_\sigma(n) = \max\{\sigma(T_{s_1, \ldots, s_n}); s_1, \ldots, s_n \in \mathbb{R}\} .$$

(See (1.22).) We show

Theorem G. Const $\sqrt{\log(n+1)} \leqq ex_\sigma(n) \leqq$ Const $\sqrt{\log(n+1)}$ $(n \geqq 1)$.

The first inequality is shown in §3.4. We prove the second inequality. For a positive integer n, F_n denotes the totality of sets $E \subset \mathbb{C}$ such that $E \subset \bigcup_{k=-\infty}^{\infty}[I_0 + ik/n]$, E has a finite number of components and their projections to I_0 are mutually disjoint, For $E \in F_n$, we define a function $A_E(x)$ on $\mathbb{R}$ by $x + iA_E(x) \in E$ $(x \in pr(E))$ and $A_E(x) = 0$ $(x \notin pr(E))$, where $pr(E)$ is the projection of E to I_0. We define a kernel by

$$T_E(x,y) = 1/\{(x-y) + i(A_E(x) - A_E(y))\} .$$

Here are three lemmas necessary for the proof.

Lemma 4.1. Let $E \in F_n$ and let W_1, W_2 be two disjoint subsets of $pr(E)$ such that $A_E(x) \geqq 0$ on W_1 and $A_E(x) \leqq 0$ on W_2. Then, for any $f \in L^2$,

$$(4.2) \qquad \int_{W_1} |T_E(\chi_{W_2} f)(x)|^2 \, dx \leqq \text{Const} \, \|\chi_{W_2} f\|_2^2 .$$

Proof. We define an operator T'_E by

$$g \to \int_{-\infty}^{\infty} g(y)/\{(x-y) - i\, A(y)\} \, dy.$$

Let T''_E denote the adjoint operator of T'_E. Then we have

$$|T''g(x)| \leq H^*g(x) + \text{Const}\, \mathcal{M}g(x),$$

which shows that $\|T''_E\|_{p,p} \leq C_p$ $(p > 1)$. Hence $\|T'_E\|_{p,p} \leq C_p$ $(p > 1)$. Since $A_E(x) - A_E(y) \geq A_E(x) \geq 0$ $(x \in W_1,\ y \in W_2)$, we have, in the same manner as in the proof of (2.9),

$$|T_E(\chi_{W_2} f)(x)| \leq \text{Const}\,\{\mathcal{M}(T'_E f)(x) + \|T'_E\|_{4/3,4/3}\, \mathcal{M}(|\chi_{W_2} f|^{4/3})(x)^{3/4}\} \quad (x \in W_1),$$

which gives (4.2).

Put

$$\xi(n) = \sup\{\frac{1}{|pr(E)|}\xi(E,f);\ E \in \mathcal{F}_n,\ f \in L^\infty_{real},\ 0 \leq f \leq 1\},$$

where

$$\xi(E,f) = \int_{pr(E)} |T_E(\chi_{pr(E)} f)(x)|^2 f(x)\,dx.$$

Lemma 4.2. For any $n \geq 1$, $\xi(n) < \infty$.

Proof. For $E \in \mathcal{F}_1$, we put $G = pr(E)$, $G'_\mu = pr(E \cap \{\text{Im } z = \mu\})$ $(\mu = 0, \pm 1, \ldots)$. Then, for any $f \in L_{real}$, $0 \leq f \leq 1$,

$$|T_E(\chi_G f)(x) - \int_{G'_k} \frac{f(y)}{x-y}\,dy - \sum_{\mu=-\infty}^{\infty} \frac{1}{1+i(k-\mu)} \int_{G'_\mu} f(y)\,dy| \leq \text{Const} \sum_{\mu=-\infty}^{\infty} \frac{|G'_\mu|}{(k-\mu)^2+1} \quad (x \in G'_k,\ k = 0, \pm 1, \ldots).$$

Hence

$$\xi(E,f) \leq \text{Const}\,\{\sum_{k=-\infty}^{\infty} \int_{G'_k} |H(\chi_{G'_k} f)(x)|^2\,dx + \sum_{k=-\infty}^{\infty} |G'_k|\,|\sum_{\mu=-\infty}^{\infty} \frac{1}{1+i(k-\mu)} \int_{G'_\mu} f(y)\,dy|^2 + \sum_{k=-\infty}^{\infty} |G'_k| \sum_{\mu=-\infty}^{\infty} \frac{|G'_\mu|}{(k-\mu)^2+1}\}$$
$$\leq \text{Const} \sum_{k=-\infty}^{\infty} \{|G_k| + |\int_{G'_\mu} f(y)\,dy|^2 + |G'_k|\} \leq \text{Const}\,|G|,$$

which shows that $\xi(1) \leq \text{Const}$.

For $E \in \mathcal{F}_n$, $f \in L^\infty_{real}$, $0 \leq f \leq 1$, we put $G' = [n\, pr(E)]$,

$$E_k = \{z - k;\ z \in [n\,E],\ k \leq \text{Re } z < k+1\},$$

$$f_k(x) = f(\frac{x+k}{n}) \qquad (0 \leq k \leq n-1).$$

Then

$$\xi(E,f) = \frac{1}{n} \int_{G'} \left| \int_{G'} \frac{f(y/n)\, dy}{(x-y) + i(n\, A_E(\frac{x}{n}) - n\, A_E(\frac{y}{n}))} \right|^2 f(\tfrac{x}{n})\, dx$$
$$\leq \frac{1}{n} \{ \sum_{k=1}^{n} \xi(E_k, f_k) + C_n \}.$$

Since $E_k \in \mathcal{F}_1$, $f_k \in L^\infty_{real}$, $0 \leq f \leq 1$, we have $\xi(n) < \infty$. Q.E.D.

The following Lemma is analogous to Lemma 3.4.

Lemma 4.3. $\xi(2^{2n}) \leq 2\, \xi(2^n) + \text{Const}\, \xi(2^{2n})^{1/2}$ $(n \geq 1)$.

Proof. For $E \in \mathcal{F}_{2^{2n}}$, we define

$$F = \bigcup_{k=-\infty}^{\infty} (\text{the projection of } E \cap \{(k-1)2^{-n} \leq \text{Im } z < k\, 2^{-n}\}$$
$$\text{to the line Im } z = (k-1)2^{-n}).$$

Then $F \in \mathcal{F}_{2^n}$. Let

$$G = \text{pr}(E),$$

$$G_j = \text{pr}(E \cap \{(j-1)2^{-n} \leq \text{Re } z < j\, 2^{-n}\}),$$

$$G_{j,k} = \text{pr}(E \cap \{(j-1)2^{-n} \leq \text{Re } z < j\, 2^{-n},\ (k-1)2^{-n} \leq \text{Im } z < k\, 2^{-n}\})$$
$$(j = 1, \ldots, 2^n,\ k = 0, \pm 1, \ldots).$$

We have, for $f \in L_{real}$, $0 \leq f \leq 1$,

$$\xi(E,f) = \xi(F,f)$$
$$+ \int_G (T_E - T_F)(\chi_G f)(x)\, \overline{T_E(\chi_G f)(x)}\, f(x)dx$$
$$+ \int_G T_F(\chi_G f)(x)\, \overline{(T_E - T_F)(\chi_G f)(x)}\, f(x)dx$$
$$= \xi(F,f) + L^{(1)} + L^{(2)}$$

and

$$L^{(1)} = \sum_{j=1}^{2^n} \int_{G_j} (T_E - T_F)(\chi_{G_j} f)(x)\, \overline{T_E(\chi_{G_j} f)(x)}\, f(x)dx$$
$$+ \sum_{j=1}^{2^n} \int_{G_j} (T_E - T_F)(\chi_{G_j} f)(x)\, \overline{T_E(\chi_{G-G_j} f)(x)}\, f(x)dx$$

$$+ \sum_{j=1}^{2^n} \int_{G-G_j} (T_E - T_F)(\chi_{G_j} f)(x) \overline{T_E(\chi_G f)(x)} f(x)dx$$

$$= L_1 + L_2 + L_3.$$

For $1 \leqq j \leqq 2^n$, there exists $E_j' \in F_{2^n}$, $F_j' \in F_1$ and $f_j \in L^\infty_{real}$, $0 \leqq f_j \leqq 1$ such that $|E_j'| = |F_j'| = 2^n|G_j|$,

$$\int_{G_j} |T_E(\chi_{G_j} f)(x)|^2 f(x)dx = 2^{-n} \xi(E_j',f_j)$$

$$\int_{G_j} |T_F(\chi_{G_j} f)(x)|^2 f(x)dx = 2^{-n} \xi(F_j',f_j).$$

Hence

$$|L_1| \leqq \sum_{j=1}^{2^n} \int_{G_j} |T_E(\chi_{G_j} f)(x)|^2 f(x)dx$$

$$+ \sum_{j=1}^{2^n} \{\int_{G_j} |T_F(\chi_{G_j} f)(x)|^2 f(x)dx\}^{1/2} \{\int_{G_j} |T_E(\chi_{G_j} f)(x)|^2 f(x)\, dx\}^{1/2}$$

$$= 2^{-n} \sum_{j=1}^{2^n} \xi(E_j',f_j) + 2^{-n} \sum_{j=1}^{2^n} \xi(F_j',f_j)^{1/2} \xi(E_j',f_j)^{1/2}$$

$$\leqq \xi(2^n) \sum_{j=1}^{2^n} |G_j| + \xi(1)^{1/2} \xi(2^n)^{1/2} \sum_{j=1}^{2^n} |G_j|$$

$$= |G| \{\xi(2^n) + \xi(1)^{1/2} \xi(2^n)^{1/2}\} .$$

Since $F_{2^n} \subset F_{2^{2n}}$, we have $\xi(2^n) \leqq \xi(2^{2n})$. Thus Lemma 4.2 yields that

$$|L_1| \leqq |G| \{\xi(2^n) + \text{Const}\ \xi(2^{2n})^{1/2}\} .$$

Let $x_j = (j-1)2^{-n}$ and $\tilde{G}_j = G_{j-1} \cup G_j \cup G_{j+1}$ $(1 \leqq j \leqq 2^n)$, where $G_0 = \emptyset$. Then, for any $x \in G - \tilde{G}_j$,

$$|(T_E - T_F)(\chi_{G_j} f)(x)| \leqq \text{Const} \int_{G_j} \frac{2^{-n}}{(x-y)^2 + 2^{-2n}} f(y)dy$$

$$\leqq \text{Const} \frac{2^{-n}|G_j|}{(x-x_j)^2 + 2^{-2n}} .$$

For any $g \in L^2$, we have

$$\left|\int_{-\infty}^{\infty} \sum_{j=1}^{2^n} \frac{2^{-n}|G_j|}{(x-x_j)^2 + 2^{-2n}} \, g(x)dx\right|$$

$$\leq \text{Const} \sum_{j=1}^{2^n} \int_{G_j} \left\{\int_{-\infty}^{\infty} \frac{2^{-n}}{(x-y)^2 + 2^{-2n}} |g(x)|dx\right\} dy$$

$$\leq \text{Const} \int_G Mg(x)\,dx \leq \text{Const} \sqrt{|G|}\; \|g\|_2 ,$$

which shows that

$$\left\| \sum_{j=1}^{2^n} \frac{2^{-n}|G_j|}{(\cdot-x_j)^2 + 2^{-2n}} \right\|_2 \leq \text{Const} \sqrt{|G|} .$$

Thus

$$|L_3| \leq \sum_{j=1}^{2^n} \left\{\int_{G-\tilde{G}_j} + \int_{\tilde{G}_j-G_j}\right\} |(T_E-T_F)(\chi_{G_j} f)(x) T_E(\chi_G f)(x)| \; f(x)dx$$

$$\leq \text{Const} \sum_{j=1}^{2^n} \int_{G-\tilde{G}_j} \frac{2^{-n}|G_j|}{(x-x_j)^2 + 2^{-2n}} |T_E(\chi_G f)(x)| f(x)dx$$

$$+ \; 2 \sum_{j=1}^{2^n} \int_{\tilde{G}_j-G_j} \left(\int_{G_j} \frac{dy}{|x-y|}\right) |T_E(\chi_G f)(x)| \; f(x)dx$$

$$\leq \text{Const} \int_G \left\{ \sum_{j=1}^{2^n} \frac{2^{-n}|G_j|}{(x-x_j)^2 + 2^{-2n}} \right\} |T_E(\chi_G f)(x)| f(x)dx$$

$$+ \; 2 \sum_{j=1}^{2^n} \left\{\int_{\tilde{G}_j-G_j} \left(\int_{G_j} \frac{dy}{|x-y|}\right)^2 dx\right\}^{1/2} \left\{\int_{\tilde{G}_j} |T_E(\chi_G f)(x)|^2 \; f(x)dx\right\}^{1/2}$$

$$\leq \text{Const} \sqrt{|G|}\; \xi(E,f)^{1/2} + \text{Const} \sum_{j=1}^{2^n} \sqrt{|G_j|} \left\{\int_{\tilde{G}_j} |T_E(\chi_G f)(x)|^2 f(x)dx\right\}^{1/2}$$

$$\leq \text{Const} \sqrt{|G|}\; \xi(E,f)^{1/2} + \text{Const} \sqrt{|G|}\; \xi(E,f)^{1/2}$$

$$\leq \text{Const} \; |G| \; \xi(2^{2n})^{1/2}.$$

Let $\hat{G}_{j,k} = G_{j,k-1} \cup G_{j,k} \cup G_{j,k+1}$ $(j = 1, \ldots, 2^n, \; k = 0, \pm 1, \ldots)$.

Then

$$|L_2| \leq \left| \sum_{j=1}^{2^n} \sum_{k=-\infty}^{\infty} \int_{G_{j,k}} (T_E-T_F)(\chi_{G_{j,k}} f)(x) \; \overline{T_E(\chi_{G-\tilde{G}_j} f)(x)} \; f(x)dx \right|$$

$$+ \left| \sum_{j=1}^{2^n} \sum_{k=-\infty}^{\infty} \int_{G_{j,k}} (T_E-T_F)(\chi_{G_{j,k-1} \cup G_{j,k+1}} f)(x) \; \overline{T_E(\chi_{G-\tilde{G}_j} f)(x)} \; f(x)dx \right|$$

$$+ \Big| \sum_{j=1}^{2^n} \sum_{k=-\infty}^{\infty} \int_{G_{j,k}} (T_E - T_F)(\chi_{G_j - \hat{G}_{j,k}} f)(x) \, \overline{T_E(\chi_{G-\tilde{G}_j} f)(x)} \, f(x)dx \Big|$$

$$+ \Big| \sum_{j=1}^{2^n} \int_{G_j} (T_E - T_F)(\chi_{G_j} f)(x) \, \overline{T_E(\chi_{\tilde{G}_j - G_j} f)(x)} \, f(x)dx \Big|$$

$$= L_{21} + L_{22} + L_{23} + L_{24}.$$

We have

$$|L_{24}| \le 2 \sum_{j=1}^{2^n} \{\int_{G_j} |(T_E - T_F)(\chi_{G_j} f)(x)|^2 \, f(x)dx\}^{1/2}$$

$$\times \{\int_{G_j} (\int_{\tilde{G}_j - G_j} \frac{dy}{|x-y|})^2 \, dx\}^{1/2} \le \text{Const } |G| \, \xi(2^{2n})^{1/2}.$$

Note that $\xi(2^{2n}) \ge \xi(1) \ge$ Const. Since

$$|(T_E - T_F)(\chi_{G_j - \hat{G}_{j,k}} f)(x)|$$

$$\le \text{Const} \sum_{\mu=-\infty}^{\infty} \frac{2^n |G_{j,\mu}|}{(k-\mu)^2 + 1} \le \text{Const} \quad (x \in G_{j,k}),$$

we have

$$|L_{23}| \le \text{Const} \sum_{j=1}^{2^n} \sum_{k=-\infty}^{\infty} \int_{G_{j,k}} |T_E(\chi_{G-\tilde{G}_j} f)(x)| \, f(x)dx$$

$$= \text{Const} \sum_{k=1}^{2^n} \int_{G_j} |T_E \{(\chi_G - \chi_{\tilde{G}_j - G_j} - \chi_{G_j}) f\}(x)| \, f(x)dx$$

$$\le \text{Const } |G| \, \{\xi(2^{2n})^{1/2} + 1\} \le \text{Const } |G| \, \xi(2^{2n})^{1/2}.$$

Lemma 4.1 shows that

$$|L_{22}| \le \{ \sum_{j=1}^{2^n} \sum_{k=-\infty}^{\infty} \int_{G_{j,k}} |(T_E - T_F)(\chi_{G_{j,k-1} \cup G_{j,k+1}} f)(x)|^2 \, dx\}^{1/2}$$

$$\times \{ \sum_{j=1}^{2^n} \int_{G_j} |T_E(\chi_{G-\tilde{G}_j} f)(x)|^2 \, f(x)dx\}^{1/2}$$

$$\le \text{Const } \{ \sum_{j=1}^{2^n} \sum_{k=-\infty}^{\infty} \int_{G_{j,k-1} \cup G_{j,k+1}} f(x)^2 \, dx\}^{1/2}$$

$$\times \{\xi(E,f) + \sum_{j=1}^{2^n} \int_{\tilde{G}_j} |T_E(\chi_{\tilde{G}_j} f)(x)|^2 f(x)dx\}^{1/2}$$

$$\leq \text{Const } |G| \ \xi(2^{2n})^{1/2}.$$

Since $(T_E - T_F)(x,y)$ is anti-symmetric, we have

$$\int_{G_{j,k}} (T_E - T_F)(\chi_{G_{j,k}} f)(x) f(x) dx = 0.$$

For $G_{j,k} \neq \emptyset$, we choose a point $x_{j,k}$ on $G_{j,k}$. Then

$$|L_{21}| = \Big| \sum_{j,k;G_{j,k}\neq\emptyset} \int_{G_{j,k}} (T_E - T_F)(\chi_{G_{j,k}} f)(x) \times \overline{\{T_E(\chi_{G-\tilde{G}_j} f)(x) - T_E(\chi_{G-\tilde{G}_j} f)(x_{j,k})\}}\, f(x) dx$$

$$\leq \text{Const} \sum_{j,k;G_{j,k}\neq\emptyset} \int_{G_{j,k}} |(T_E - T_F)(\chi_{G_{j,k}} f)(x)| f(x) dx$$

$$\leq \text{Const } |G| \ \xi(2^{2n})^{1/2}.$$

Consequently,

$$|L^{(1)}| \leq |G| \ \{\xi(2^n) + \text{Const } \xi(2^{2n})^{1/2}\}.$$

In the same manner,

$$|L^{(2)}| \leq \sum_{j=1}^{2^n} \int_{G_j} |T_F(\chi_{G_j} f)(x) T_E(\chi_{G_j} f)(x)| f(x) dx + \text{Const } |G| \ \xi(2^{2n})^{1/2}.$$

Since the first quantity in the left hand side of the above inequality is dominated by Const $|G| \ \xi(2^{2n})^{1/2}$, we obtain $|L^{(2)}| \leq$ Const $|G| \ \xi(2^{2n})^{1/2}$. Thus

$$\xi(E,f) \leq \xi(F,f) + |G| \ \{\xi(2^n) + \text{Const } \xi(2^{2n})^{1/2}\}$$

$$\leq |G| \ \{2\ \xi(2^n) + \text{Const } \xi(2^{2n})^{1/2}\},$$

which yields the required inequality. Q.E.D.

We now prove the second inequality in Theorem G. Lemmas 4.2 and 4.3 show that

$$\xi(2^{2^n}) \leq 2\ \xi(2^{2^{n-1}}) + \text{Const } \xi(2^{2^n})^{1/2} \leq \ldots$$

$$\leq 2^n\ \xi(2^{2^0}) + \text{Const} \sum_{k=0}^{n-1} 2^k\ \xi(2^{2^{n-k}})^{1/2}$$

$$\leq \text{Const } \{2^n + \sum_{k=0}^{n-1} 2^k\ \xi(2^{2^{n-k}})^{1/2}\},$$

which yields that $\xi(2^{2^n}) \leq \text{Const } 2^n$. (See the proof of (2.43).) Let ℓ_n denote the integer satisfying $2^{\ell_n - 1} \leq n < 2^{\ell_n}$. Then

$$\xi(2^n) \leq \xi(2^{2^{\ell_n}}) \leq \text{Const } 2^{\ell_n} \leq \text{Const } n.$$

For $E \in F_n$ and $f \in L^\infty_{\text{real}}$, $0 \leq f \leq 1$, we put $F = [n2^{-\ell_n} E]$ and $g(x) = f(x\, 2^{\ell_n}/n)$. Then $F \in F_{2^{\ell_n}}$. Hence we have

$$\begin{aligned} \xi(E,f) &= \frac{2^{\ell_n}}{n} \xi(F,g) \leq \text{Const } |F|\, \xi(2^{\ell_n}) \\ &\leq \text{Const } |E|\, \ell_n \leq \text{Const } |E| \log(n+1), \end{aligned}$$

which gives that

$$(4.3) \quad \xi(n) \leq \text{Const } \log(n+1) \qquad (n \geq 1).$$

For $s_1, \ldots, s_n \in \mathbb{R}$, we put

$$E(\hat{s}_1, \ldots, \hat{s}_n) = \bigcup_{k=1}^{n} \{x + \hat{s}_k,\ \frac{k-1}{n} \leq x < \frac{k}{n}\},$$

where $\hat{s}_k$ = (the integral part of ns_k)/n. Then we have, for an interval $I \subset I_0$ and $f \in L^\infty_{\text{real}}$, $0 \leq f \leq 1$,

$$\hat{\sigma}(I, T_{s_1, \ldots, s_n}, f) \leq \text{Const } \{\hat{\sigma}(I, T_{E(\hat{s}_1, \ldots, \hat{s}_n)}, f) + |I|\}.$$

Hence (4.3) shows that $\hat{\sigma}(I, T_{s_1, \ldots, s_n}, f) \leq \text{Const } |I| \log(n+1)$. Since $T_{s_1, \ldots, s_n}(x,y) = 1/(x-y)$ $(x, y \notin I_0)$, this inequality gives $\hat{\sigma}(T_{s_1, \ldots, s_n}) \leq \text{Const } \log(n+1)$. Consequently,

$$\sigma(T_{s_1, \ldots, s_n}) \leq \text{Const } \hat{\sigma}(T_{s_1, \ldots, s_n})^{1/2} \leq \text{Const} \sqrt{\log(n+1)}.$$

Since $s_1, \ldots, s_n \in \mathbb{R}$ are arbitrary, the second inequality in Theorem G holds. This completes the proof of Theorem G.

Let $BMO(\Gamma)$ denote the Banach space of functions f on a finite union Γ of segments, modulo constants, with norm

$$\|f\|_{BMO(\Gamma)} = \sup \frac{1}{|\Gamma \cap D(z,2r)|} \int_{\Gamma \cap D(z,r)} |f(z) - (f)_{\Gamma \cap D(z,r)}|\,|dz|,$$

where $(f)_{\Gamma \cap D(z,r)}$ is the mean of f over $\Gamma \cap D(z,r)$ with respect to $|dz|$ and the supremum is taken over all $z \in \mathbb{C}$, $r > 0$. Put

$\Gamma_{s_1, \ldots, s_n} = \{(x, A_{s_1, \ldots, s_n}(x));\ x \in I_0\}$. Theorem G immediately yields

Corollary 4.4. $\text{Const}\sqrt{\log(n+1)}$

$$\leq \max\{\|H_{\Gamma_{s_1, \ldots, s_n}}\|_{L^\infty(\Gamma_{s_1, \ldots, s_n}), BMO(\Gamma_{s_1, \ldots, s_n})};\ s_1, \ldots, s_n \in \mathbb{R}\}$$

$$\leq \text{Const}\sqrt{\log(n+1)} \quad (n \geq 1),$$

where $\|H_\Gamma\|_{L^\infty(\Gamma), BMO(\Gamma)}$ is the norm of H_Γ as an operator from $L^\infty(\Gamma)$ to $BMO(\Gamma)$.

APPENDIX II. PROOF OF THEOREM B BY P. W. JONES-S. SEMMES

Quite recently, P. W. Jones-S.Semmes gave a proof of Theorem B by complex variable methods. The following note is their lecture in May, 1987. (The author expresses his thanks to P. W. Jones-S.Semmes who permitted the author to write here their proof (cf. P.W. Jones [33]). Here is a fact obtained by C. Kenig.

Lemma 4.5. Let $\Gamma = \{x + iA(x);\ x \in \mathbb{R}\}$ be a Lipschitz graph and $\Omega = \{z \in \mathbb{C};\ \mathrm{Im}\, z > A(\mathrm{Re}\, z)\}$. Then, for any $g \in L^2(\Gamma)$ having an analytic extension, say simply $g(z)$ $(z \in \Omega)$, to Ω,

$$(4.4)\quad \|g\|_{L^2(\Gamma)} \leq C_M \{\iint_\Omega |g'(z)|^2\, \mathrm{dis}(z,\Gamma)\, d\sigma(z)\}^{1/2},$$

where $d\sigma$ is the area element and C_M is a constant depending only on $M = \|A'\|_\infty$.

For $z \in \Omega$, we write $z^* = z - 2i(\mathrm{Im}\, z - A(\mathrm{Re}\, z))$. For $f \in L^2(\Gamma)$, we put $\tilde{C}f(z) = C(f\, d\zeta|_\Gamma)(z)$ $(z \in \Omega)$, i.e.,

$$\tilde{C}f(z) = \frac{1}{2\pi i} \int_\Gamma \frac{f(\zeta)}{\zeta - z}\, d\zeta.$$

For $z \in \Gamma$, we write by $\tilde{C}f(z)$ the nontangential limit of $\tilde{C}f(\zeta)$ $(\zeta \in \Omega)$ at z. For $z \in \Gamma \cup \Omega$, we have

$$\begin{aligned}\tilde{C}f(z) &= -i \int_0^\infty (\tilde{C}f)'(z + it)\,dt = \int_0^\infty (\tilde{C}f)''(z + it)\,t\, dt \\ &= \frac{1}{2\pi i} \int_0^\infty \{\int_\Gamma \frac{(\tilde{C}f)'(\zeta + it/2)t}{(z+(it/2) - \zeta)^2}\, d\zeta\}\, dt \\ &= \frac{2}{\pi i} \iint_\Omega \frac{(\tilde{C}f)'(w)}{(z - w^*)^2} \{\mathrm{Im}\, w - A(\mathrm{Re}\, w)\}\{1 + iA'(\mathrm{Re}\, w)\}\, d\sigma(w)\end{aligned}$$

Let $\{Q_k\}_{k=1}^\infty$ be a sequence of mutually disjoint cubes (with sides parallel to the coordinate axes) such that $\Omega = \bigcup_{k=1}^\infty Q_k$, $d_k/C_M \leq \ell(Q_k) \leq C_M d_k$ $(k \geq 1)$, where $\ell(Q_k)$ is the length of a side of Q_k and $d_k = \mathrm{dis}(Q_k,\Gamma)$. Then

$$\tilde{C}f(z) = \frac{2}{\pi i} \sum_{k=1}^\infty \iint_{Q_k} \frac{(\tilde{C}f)'(w)}{(z - w^*)^2} \{\mathrm{Im}\, w - A(\mathrm{Re}\, w)\}\{1 + iA'(\mathrm{Re}\, w)\}\, d\sigma(w),$$

$$\iint_{Q_k} |\mathrm{Im}\, w - A(\mathrm{Re}\, w)|\, d\sigma(w) \leq C_M\, \mathrm{dis}(Q_k,\Gamma)^3 \quad (k \geq 1).$$

Hence

$$(4.5)\quad \|\tilde{C}f\|_{L^2(\Gamma)} \leq C_M \sup \| \sum_{k=1}^{\infty} c_k(\tilde{C}f)'(z_k) \left(\frac{d_k}{z - z_k^*} \right)^2 \|_{L^2(\Gamma)},$$

where the supremum is taken over all sequences $\{c_k\}_{k=1}^{\infty}$, $\{z_k\}_{k=1}^{\infty}$ such that $|c_k| \leq d_k$, $z_k \in Q_k$ $(k \geq 1)$. Let M_2 denote the 2-dimension maximal operator. For two sequences $\{c_k\}_{k=1}^{\infty}$, $\{z_k\}_{k=1}^{\infty}$ satisfying the above condition, we define a function $h(\zeta)$ on $\mathbb{C}$ by

$$h(\zeta) = c_k(\tilde{C}f)'(z_k)\, d_k^{-1/2} \quad (\zeta \in Q_k,\ k \geq 1),\quad h(\zeta) = 0 \quad (\zeta \in \mathbb{C} - \Omega).$$

Then Lemma 4.5 shows that

$$(4.6)\quad \| \sum_{k=1}^{\infty} c_k(\tilde{C}f)'(z_k) \left(\frac{d_k}{z - z_k^*} \right)^2 \|_{L^2(\Gamma)}$$

$$\leq C_M \{\iint_{\Omega} | \sum_{k=1}^{\infty} c_k(\tilde{C}f)'(z_k) \frac{d_k^2}{(z - z_k^*)^3} |^2 \operatorname{dis}(z,\Gamma)\, d\sigma(z)\}^{1/2}$$

$$\leq C_M \{\iint_{\Omega} |\iint_{\mathbb{C}} h(\zeta) \frac{|\operatorname{Im}\zeta|^{1/2} |\operatorname{Im} z|^{1/2}}{|z - \zeta^*|^3} d\sigma(\zeta)|^2 d\sigma(z)\}^{1/2}$$

$$\leq C_M \{\iint_{\Omega} M_2 h(z)^2\, d\sigma(z)\}^{1/2} \leq C_M \{\iint_{\mathbb{C}} |h(z)|^2 d\sigma(z)\}^{1/2}$$

$$\leq C_M \{ \sum_{k=1}^{\infty} |c_k(\tilde{C}f)'(z_k)|^2 d_k\}^{1/2} \leq C_M \{ \sum_{k=1}^{\infty} |(\tilde{C}f)'(z_k)|^2 d_k^3\}^{1/2}.$$

Let G denote the totality of sequences $\{\alpha_k\}_{k=1}^{\infty}$ such that $\sum_{k=1}^{\infty} |\alpha_k|^2 d_k \leq 1$. Then

$$(4.7)\quad \{ \sum_{k=1}^{\infty} |(\tilde{C}f)'(z_k)|^2 d_k^3\}^{1/2}$$

$$= \sup \{| \sum_{k=1}^{\infty} (\tilde{C}f)'(z_k)\, \alpha_k\, d_k^2 |,\ \{\alpha_k\}_{k=1}^{\infty} \in G\}.$$

Lemma 4.4 shows that, for any $\{\alpha_k\}_{k=1}^{\infty} \in G$,

$$(4.8)\quad | \sum_{k=1}^{\infty} (\tilde{C}f)'(z_k)\, \alpha_k\, d_k^2 | = \frac{1}{2\pi} |\int_{\Gamma} f(z) \sum_{k=1}^{\infty} \alpha_k \frac{d_k^2}{(z - z_k)^2} dz |$$

$$\leq \frac{1}{2\pi} \|f\|_{L^2(\Gamma)} \| \sum_{k=1}^{\infty} \alpha_k \frac{d_k^2}{(z - z_k)^2} \|_{L^2(\Gamma)}$$

$$\leq C_M \|f\|_{L^2(\Gamma)} \{ \iint_{\mathbb{C}-\Omega} | \sum_{k=1}^{\infty} \alpha_k \frac{d_k^2}{(z - z_k)^3} |^2 \operatorname{dis}(z,\Gamma)\, d\sigma(z)\}^{1/2}$$

$$\leq C_M \|f\|_{L^2(\Gamma)} \{\sum_{k=1}^{\infty} |\alpha_k|^2 d_k\}^{1/2} \leq C_M \|f\|_{L^2(\Gamma)} .$$

Thus (4.5)-(4.8) show that $\|\tilde{C}f\|_{L^2(\Gamma)} \leq C_M \|f\|_{L^2(\Gamma)}$, which yields Theorem B.

The proof of Lemma 4.5 by P.W.Jones-S.Semmes is as follows. Put $A = \|g\|^2_{L^2(\Gamma)}$, $B = \iint_\Omega |g'(z)|^2 \operatorname{dis}(z,\Gamma)\, d\sigma(z)$. Let $\Phi(z)$ be a conformal one to one mapping form the upper half plane U to Ω. Then

$$A = \int_{-\infty}^{\infty} |g \circ \Phi(x)|^2 \, |\Phi'(x)| \, dx$$

$$B = \iint_U |g' \circ \Phi(z)|^2 \operatorname{dis}(\Phi(z),\Gamma) \, |\Phi'(z)|^2 \, d\sigma(z)$$

Koebe's 1/4-theorem shows that $\operatorname{dis}(\Phi(z),\Gamma) \geq |\Phi'(z)|(\operatorname{Im} z)/4$ $(z \in U)$, and hence

$$\iint_U |g' \circ \Phi(z)|^2 \, |\Phi'(z)|^3 \, y \, d\sigma(z) \leq \text{Const } B.$$

Since $|\arg \Phi'(x)| \leq \pi/2 - (1/M)$ $(x \in \mathbb{R})$, Green's formula shows that

$$\begin{aligned} A &\leq C_M \left|\int_{-\infty}^{\infty} |g \circ \Phi(x)|^2 \, \Phi'(x) \, dx\right| \\ &= C_M \left|\iint_U \Delta(|g \circ \Phi(z)|^2 \, \Phi'(z)) \, y \, d\sigma(z)\right| \\ &\leq C_M \{\iint_U |g' \circ \Phi(z)|^2 \, |\Phi'(z)|^3 \, y \, d\sigma(z) \\ &\quad + \iint_U |(g' \circ \Phi)(z)(g \circ \Phi)(z)| |\Phi'(z)|^2 \, |\Phi''(z)| \, y \, d\sigma(z)\} \\ &\leq C_M[B + B^{1/2}\{\iint_U |g \circ \Phi(z)|^2 |\Phi'(z)| \left|\frac{\Phi''(z)}{\Phi'(z)}\right|^2 y \, d\sigma(z)\}^{1/2}]. \end{aligned}$$

We can write $\Phi'(z) = e^{V(z)}$ with an analytic function $V(z)$ in U. Then $V \in BMO$ since $\operatorname{Im} V \in L^\infty$. Thus

$$|\Phi''(z)/\Phi'(z)|^2 \, y \, dy \, dx \quad (= |V'(z)|^2 \, y \, dx \, dy)$$

is a Carleson measure in U. Since $g \circ \Phi(z) \, e^{V(z)/2}$ is analytic in U,

$$\{\iint_U |g \circ \Phi(z)|^2 |\Phi'(z)| \left|\frac{\Phi''(z)}{\Phi'(z)}\right|^2 y \, d\sigma(z)\}^{1/2} \leq C_M \, A^{1/2}.$$

Thus $A \leq C_M(B + B^{1/2}A^{1/2})$, which yields the required inequality.

REFERENCES

[1] A.S.Besicovitch, On the fundamental geometrical properties of linearly measurable plane sets of points (III), Math. Ann., 116 (1939), 349-357.

[2] D.L.Burkholder and R.R.Gundy, Distribution function inequalities for the area integral, Studia Math., 44 (1972), 527-544.

[3] A.P.Calderón, Commutators of singular integral operators, Proc. Nat. Acad. Sci. USA, 53 (1965), 1092-1099.

[4] A.P.Calderón, Cauchy integrals on Lipschitz curves and related operators, Proc. Nat. Acad. Sci. USA, 74 (1977), 1324-1327.

[5] A.P.Calderón, Commutators, singular integrals on Lipschitz curves and applications, ICM Helsinki (1978).

[6] A.P.Calderón, C.P.Calderón, E.B.Fabes, M.Jodeit Jr. and N.M.Riviere, Applications of the Cauchy integral along Lipschitz curves, Bull. Amer. Math. Soc., 84 (1978), 287-290.

[7] R.R.Coifman, A.McIntosh and Y.Meyer, L'intégrale de Cauchy définit un opérateur borné sur L^2 pour les courbes lipschitziennes, Ann. of Math., 116 (1982), 361-387.

[8] R.R.Coifman and Y.Meyer, On commutators of singular integrals and bilinear singular integrals, Trans. Amer. Math. Soc., 212 (1975), 315-331.

[9] R.R.Coifman and Y.Meyer, Commutateurs d'intégrales singulières et opérateurs multilinéaires, Ann. Inst. Fourier Grenoble, 28 (1978), 177-202.

[10] R.R.Coifman and Y.Meyer, "au-delà des opérateurs pseudo-différentiels", Astérisque 57, Soc. Math. France, 1978.

[11] R.R.Coifman and Y.Meyer, Le théorème de Calderón par les méthodes de variable réelle, C.R. Acad. Sc. Paris, 289 (1979), 425-428.

[12] R.R.Coifman and Y.Meyer, Non-linear harmonic analysis, operator theory and P.D.E., in Beijing lectures in harmonic analysis (edited by E.M.Stein), 3-45, Princeton Univ. Press, 1986.

[13] R.R.Coifman, Y.Meyer and E.M.Stein, Un nouvel espace fonctionnel adapté à l'étude des opérateurs définis par des intégrales singulières, in Harmonic Analysis, Cortona, 1-15, Lecture Notes in Math., 992, Springer-Verlag, 1982.

[14] R.R.Coifman, Y.Meyer and E.M.Stein, Some new function spaces and their applications to analysis, J. Funct. Anal., 62 (1985), 304-335.

[15] R.R.Coifman, R.Rochberg and G.Weiss, Factorization theorems for Hardy spaces in several variables, Ann. of Math., 103 (1976), 611-635.

[16] M.Cotlar, A unified theory of Hilbert transforms and ergodic theory, Rev. Mat. Cuyana, 1 (1955), 105-167.

[17] G.David, Opérateurs intégraux singuliers sur certaines courbes du plan complexe, Ann. Sci. Ec. Norm. Sup., 17 (1984), 157-189.

[18] G.David, Noyau de Cauchy et opérateurs de Calderón-Zygmund, Thèse d'Etat, Orsay, 3193 (1986).

[19] G.David and J.L.Journé, A boundedness criterion for generalized Calderón-Zygmund operators, Ann. of Math., 120 (1984), 371-397.

[20] G.David, J.L.Journé and S.Semmes, Opérateurs de Calderón-Zygmund, fonctions para-accrétives et interpolation, Rev. Mat. Iberoamericana, 1-4 (1985), 1-56.

[21] A.M.Davie, Analytic capacity and approximation problems, Trans. Amer. Math. Soc., 171 (1972), 409-444.

[22] A.M.Davie and B.Øksendal, Analytic capacity and differentiability properties of finely harmonic functions, Acta Math., 149 (1982), 127-152.

[23] A.Denjoy, Sur la continuité des fonctions analytiques singulières, Bull. Soc. Math. France, 60 (1932), 27-105.

[24] E.B.Fabes, N.M.Riviere and M.Jodeit Jr., Potential techniques for boundary value problems on C^1-domains, Acta Math., 141 (1978), 165-186.

[25] C.Fefferman, Recent progress in classical Fourier analysis, ICM Vancouver (1974).

[26] R.Fefferman, R.R.Gundy, M.Silverstein and E.M.Stein, Inequalities for ratios of functionals of harmonic functions, Proc. Nat. Acad. Sci. USA, 79 (1982), 7958-7960.

[27] C.Fefferman and E.M.Stein, H^p spaces of several variables, Acta Math., 129 (1972), 137-193.

[28] J.Garnett, Positive length but zero analytic capacity, Proc Amer. Math. Soc., 21 (1970), 696-699.

[29] J.Garnett, "Analytic Capacity and Measure", Lecture Notes in Math., 297 Springer-Verlag, 1972.

[30] T.E.Harris, "The theory of branching processes", Springer-Verlag, 1963.

[31] L.D.Ivanov, On sets of analytic capacity zero, in Linear and Complex Analysis-199 Research Problems, 498-501, Lecture Notes in Math., 1043, Springer-Verlag, 1984.

[32] F.John and L.Nirenberg, On functions of bounded mean oscillation, Comm. Pure Appl. Math., 14 (1961), 415-426.

[33] P.W.Jones, in preparation.

[34] P.W.Jones and T.Murai, Positive analytic capacity but zero Buffon needle probability, to appear.

[35] J.L.Journé, "Calderón-Zygmund Operators, Pseudo-Differential Operators and the Cauchy Integral of Calderón", Lecture Notes in Math., 994, Springer-Velrag, 1983.

[36] P.G.Lemarie, Continuité sur les espaces de Besov des opérateurs définis par des intégrales singulières, Ann. Inst. Fourier Grenoble, 35 (1985), 175-187.

[37] D.E.Marshall, Removable sets for bounded analytic functions, in Linear and Complex Analysis-199 Research Problems, 485-490, Lecture Notes in Math., 1043, Springer-Verlag, 1984.

[38] P.Mattila, A class of sets with positive length and zero analytic capacity, Ann. Acad. Sci. Fenn. Ser. A I 10 (1985), 387-395.

[39] P.Mattila, Smooth maps, null-sets for integralgeometric measure and analytic capacity, Ann. of Math., 123 (1986), 303-309.

[40] A.McIntosh and Y.Meyer, Algèbres d'opérateurs définis par des intégrales singulières, C.R. Acad. Sc. Paris, 301 (1985), 395-397.

[41] Y.Meyer, Le lemme de Cotlar et Stein et la continuité L^2 des opérateurs définis par des intégrales singulières, in Colloque en l'honneur de L.Schwartz, 115-126, Astérisque 131, Soc. Math. France, 1985.

[42] T.Murai, Boundedness of singular integral operators of Calderón type (IV), Hiroshima Math. J., 14 (1985), 511-525.

[43] T.Murai, Boundedness of singular integral operators of Calderón type (V), Adv. in Math., 59 (1986), 71-81.

[44] T.Murai, Boundedness of singular integral operators of Calderón type VI, Nagoya Math. J., 102 (1986), 127-133.

[45] T.Murai, Proof of the boundedness of commutators by perturbation, Bull. London Math. Soc., 18 (1986), 383-388.

[46] T.Murai, Construction of H^1 functions concerning the estimate of analytic capacity, Bull. London Math. Soc., 19 (1986), 154-160.

[47] T.Murai, Comparison between analytic capacity and the Buffon needle probability, Trans. Amer. Math. Soc., to appear.

[48] T.Murai and A.Uchiyama, Good λ inequalities for the area integral and the nontangential maximal function, Studia Math., 81 (1986), 251-262.

[49] L.A.Santaló, "Introduction to integral geometry", Harmann, 1953.

[50] E.M.Stein, "Singular Integrals and Differentiability Properties of Functions", Princeton Univ. Press, 1970.

[51] P.Tchamitchian, La normede l'opérateur de Cauchy sur les courbes lipschitziennes, in preparation.

[52] A.G.Vitushkin, Example of a set of positive length but zero analytic capacity, Dokl. Akad. Nauk. SSSR, 127 (1959), 246-249 (Russin).

[53] K.Yoshida, Functional Analysis Sixth Edition, Springer-Verlag, 1980.

[54] A.Zygmund, Trigonometric series I, Cambridge Univ. Press, 1959.

SUBJECT INDEX

Vol. 1145: G. Winkler, Choquet Order and Simplices. VI, 143 pages. 1985.

Vol. 1146: Séminaire d'Algèbre Paul Dubreil et Marie-Paule Malliavin. Proceedings, 1983–1984. Edité par M.-P. Malliavin. IV, 420 pages. 1985.

Vol. 1147: M. Wschebor, Surfaces Aléatoires. VII, 111 pages. 1985.

Vol. 1148: Mark A. Kon, Probability Distributions in Quantum Statistical Mechanics. V, 121 pages. 1985.

Vol. 1149: Universal Algebra and Lattice Theory. Proceedings, 1984. Edited by S. D. Comer. VI, 282 pages. 1985.

Vol. 1150: B. Kawohl, Rearrangements and Convexity of Level Sets in PDE. V, 136 pages. 1985.

Vol 1151: Ordinary and Partial Differential Equations. Proceedings, 1984. Edited by B.D. Sleeman and R.J. Jarvis. XIV, 357 pages. 1985.

Vol. 1152: H. Widom, Asymptotic Expansions for Pseudodifferential Operators on Bounded Domains. V, 150 pages. 1985.

Vol. 1153: Probability in Banach Spaces V. Proceedings, 1984. Edited by A. Beck, R. Dudley, M. Hahn, J. Kuelbs and M. Marcus. VI, 457 pages. 1985.

Vol. 1154: D.S. Naidu, A.K. Rao, Singular Pertubation Analysis of Discrete Control Systems. IX, 195 pages. 1985.

Vol. 1155: Stability Problems for Stochastic Models. Proceedings, 1984. Edited by V.V. Kalashnikov and V.M. Zolotarev. VI, 447 pages. 1985.

Vol. 1156: Global Differential Geometry and Global Analysis 1984. Proceedings, 1984. Edited by D. Ferus, R.B. Gardner, S. Helgason and U. Simon. V, 339 pages. 1985.

Vol. 1157: H. Levine, Classifying Immersions into $\mathbb{R}^4$ over Stable Maps of 3-Manifolds into $\mathbb{R}^2$. V, 163 pages. 1985.

Vol. 1158: Stochastic Processes – Mathematics and Physics. Proceedings, 1984. Edited by S. Albeverio, Ph. Blanchard and L. Streit. VI, 230 pages. 1986.

Vol. 1159: Schrödinger Operators, Como 1984. Seminar. Edited by S. Graffi. VIII, 272 pages. 1986.

Vol. 1160: J.-C. van der Meer, The Hamiltonian Hopf Bifurcation. VI, 115 pages. 1985.

Vol. 1161: Harmonic Mappings and Minimal Immersions, Montecatini 1984. Seminar. Edited by E. Giusti. VII, 285 pages. 1985.

Vol. 1162: S.J.L. van Eijndhoven, J. de Graaf, Trajectory Spaces, Generalized Functions and Unbounded Operators. IV, 272 pages. 1985.

Vol. 1163: Iteration Theory and its Functional Equations. Proceedings, 1984. Edited by R. Liedl, L. Reich and Gy. Targonski. VIII, 231 pages. 1985.

Vol. 1164: M. Meschiari, J.H. Rawnsley, S. Salamon, Geometry Seminar "Luigi Bianchi" II – 1984. Edited by E. Vesentini. VI, 224 pages. 1985.

Vol. 1165: Seminar on Deformations. Proceedings, 1982/84. Edited by J. Ławrynowicz. IX, 331 pages. 1985.

Vol. 1166: Banach Spaces. Proceedings, 1984. Edited by N. Kalton and E. Saab. VI, 199 pages. 1985.

Vol. 1167: Geometry and Topology. Proceedings, 1983–84. Edited by J. Alexander and J. Harer. VI, 292 pages. 1985.

Vol. 1168: S.S. Agaian, Hadamard Matrices and their Applications. III, 227 pages. 1985.

Vol. 1169: W.A. Light, E.W. Cheney, Approximation Theory in Tensor Product Spaces. VII, 157 pages. 1985.

Vol. 1170: B.S. Thomson, Real Functions. VII, 229 pages. 1985.

Vol. 1171: Polynômes Orthogonaux et Applications. Proceedings, 1984. Edité par C. Brezinski, A. Draux, A.P. Magnus, P. Maroni et A. Ronveaux. XXXVII, 584 pages. 1985.

Vol. 1172: Algebraic Topology, Göttingen 1984. Proceedings. Edited by L. Smith. VI, 209 pages. 1985.

Vol. 1173: H. Delfs, M. Knebusch, Locally Semialgebraic Spaces. XVI, 329 pages. 1985.

Vol. 1174: Categories in Continuum Physics, Buffalo 1982. Seminar. Edited by F.W. Lawvere and S.H. Schanuel. V, 126 pages. 1986.

Vol. 1175: K. Mathiak, Valuations of Skew Fields and Projective Hjelmslev Spaces. VII, 116 pages. 1986.

Vol. 1176: R.R. Bruner, J.P. May, J.E. McClure, M. Steinberger, H_∞ Ring Spectra and their Applications. VII, 388 pages. 1986.

Vol. 1177: Representation Theory I. Finite Dimensional Algebras. Proceedings, 1984. Edited by V. Dlab, P. Gabriel and G. Michler. XV, 340 pages. 1986.

Vol. 1178: Representation Theory II. Groups and Orders. Proceedings, 1984. Edited by V. Dlab, P. Gabriel and G. Michler. XV, 370 pages. 1986.

Vol. 1179: Shi J.-Y. The Kazhdan-Lusztig Cells in Certain Affine Weyl Groups. X, 307 pages. 1986.

Vol. 1180: R. Carmona, H. Kesten, J.B. Walsh, École d'Été de Probabilités de Saint-Flour XIV – 1984. Édité par P.L. Hennequin. X, 438 pages. 1986.

Vol. 1181: Buildings and the Geometry of Diagrams, Como 1984. Seminar. Edited by L. Rosati. VII, 277 pages. 1986.

Vol. 1182: S. Shelah, Around Classification Theory of Models. VII, 279 pages. 1986.

Vol. 1183: Algebra, Algebraic Topology and their Interactions. Proceedings, 1983. Edited by J.-E. Roos. XI, 396 pages. 1986.

Vol. 1184: W. Arendt, A. Grabosch, G. Greiner, U. Groh, H.P. Lotz, U. Moustakas, R. Nagel, F. Neubrander, U. Schlotterbeck, One-parameter Semigroups of Positive Operators. Edited by R. Nagel. X, 460 pages. 1986.

Vol. 1185: Group Theory, Beijing 1984. Proceedings. Edited by Tuan H.F. V, 403 pages. 1986.

Vol. 1186: Lyapunov Exponents. Proceedings, 1984. Edited by L. Arnold and V. Wihstutz. VI, 374 pages. 1986.

Vol. 1187: Y. Diers, Categories of Boolean Sheaves of Simple Algebras. VI, 168 pages. 1986.

Vol. 1188: Fonctions de Plusieurs Variables Complexes V. Séminaire, 1979–85. Edité par François Norguet. VI, 306 pages. 1986.

Vol. 1189: J. Lukeš, J. Malý, L. Zajíček, Fine Topology Methods in Real Analysis and Potential Theory. X, 472 pages. 1986.

Vol. 1190: Optimization and Related Fields. Proceedings, 1984. Edited by R. Conti, E. De Giorgi and F. Giannessi. VIII, 419 pages. 1986.

Vol. 1191: A.R. Its, V.Yu. Novokshenov, The Isomonodromic Deformation Method in the Theory of Painlevé Equations. IV, 313 pages. 1986.

Vol. 1192: Equadiff 6. Proceedings, 1985. Edited by J. Vosmansky and M. Zlámal. XXIII, 404 pages. 1986.

Vol. 1193: Geometrical and Statistical Aspects of Probability in Banach Spaces. Proceedings, 1985. Edited by X. Femique, B. Heinkel, M.B. Marcus and P.A. Meyer. IV, 128 pages. 1986.

Vol. 1194: Complex Analysis and Algebraic Geometry. Proceedings, 1985. Edited by H. Grauert. VI, 235 pages. 1986.

Vol. 1195: J.M. Barbosa, A.G. Colares, Minimal Surfaces in $\mathbb{R}^3$. X, 124 pages. 1986.

Vol. 1196: E. Casas-Alvero, S. Xambó-Descamps, The Enumerative Theory of Conics after Halphen. IX, 130 pages. 1986.

Vol. 1197: Ring Theory. Proceedings, 1985. Edited by F.M.J. van Oystaeyen. V, 231 pages. 1986.

Vol. 1198: Séminaire d'Analyse, P. Lelong – P. Dolbeault – H. Skoda. Seminar 1983/84. X, 260 pages. 1986.

Vol. 1199: Analytic Theory of Continued Fractions II. Proceedings, 1985. Edited by W.J. Thron. VI, 299 pages. 1986.

Vol. 1200: V.D. Milman, G. Schechtman, Asymptotic Theory of Finite Dimensional Normed Spaces. With an Appendix by M. Gromov. VIII, 156 pages. 1986.

Vol. 1201: Curvature and Topology of Riemannian Manifolds. Proceedings, 1985. Edited by K. Shiohama, T. Sakai and T. Sunada. VII, 336 pages. 1986.

Vol. 1202: A. Dür, Möbius Functions, Incidence Algebras and Power Series Representations. XI, 134 pages. 1986.

Vol. 1203: Stochastic Processes and Their Applications. Proceedings, 1985. Edited by K. Itô and T. Hida. VI, 222 pages. 1986.

Vol. 1204: Séminaire de Probabilités XX, 1984/85. Proceedings. Edité par J. Azéma et M. Yor. V, 639 pages. 1986.

Vol. 1205: B.Z. Moroz, Analytic Arithmetic in Algebraic Number Fields. VII, 177 pages. 1986.

Vol. 1206: Probability and Analysis, Varenna (Como) 1985. Seminar. Edited by G. Letta and M. Pratelli. VIII, 280 pages. 1986.

Vol. 1207: P.H. Bérard, Spectral Geometry: Direct and Inverse Problems. With an Appendix by G. Besson. XIII, 272 pages. 1986.

Vol. 1208: S. Kaijser, J.W. Pelletier, Interpolation Functors and Duality. IV, 167 pages. 1986.

Vol. 1209: Differential Geometry, Peñíscola 1985. Proceedings. Edited by A.M. Naveira, A. Ferrández and F. Mascaró. VIII, 306 pages. 1986.

Vol. 1210: Probability Measures on Groups VIII. Proceedings, 1985. Edited by H. Heyer. X, 386 pages. 1986.

Vol. 1211: M.B. Sevryuk, Reversible Systems. V, 319 pages. 1986.

Vol. 1212: Stochastic Spatial Processes. Proceedings, 1984. Edited by P. Tautu. VIII, 311 pages. 1986.

Vol. 1213: L.G. Lewis, Jr., J.P. May, M. Steinberger, Equivariant Stable Homotopy Theory. IX, 538 pages. 1986.

Vol. 1214: Global Analysis – Studies and Applications II. Edited by Yu.G. Borisovich and Yu.E. Gliklikh. V, 275 pages. 1986.

Vol. 1215: Lectures in Probability and Statistics. Edited by G. del Pino and R. Rebolledo. V, 491 pages. 1986.

Vol. 1216: J. Kogan, Bifurcation of Extremals in Optimal Control. VIII, 106 pages. 1986.

Vol. 1217: Transformation Groups. Proceedings, 1985. Edited by S. Jackowski and K. Pawalowski. X, 396 pages. 1986.

Vol. 1218: Schrödinger Operators, Aarhus 1985. Seminar. Edited by E. Balslev. V, 222 pages. 1986.

Vol. 1219: R. Weissauer, Stabile Modulformen und Eisensteinreihen. III, 147 Seiten. 1986.

Vol. 1220: Séminaire d'Algèbre Paul Dubreil et Marie-Paule Malliavin. Proceedings, 1985. Edité par M.-P. Malliavin. IV, 200 pages. 1986.

Vol. 1221: Probability and Banach Spaces. Proceedings, 1985. Edited by J. Bastero and M. San Miguel. XI, 222 pages. 1986.

Vol. 1222: A. Katok, J.-M. Strelcyn, with the collaboration of F. Ledrappier and F. Przytycki, Invariant Manifolds, Entropy and Billiards; Smooth Maps with Singularities. VIII, 283 pages. 1986.

Vol. 1223: Differential Equations in Banach Spaces. Proceedings, 1985. Edited by A. Favini and E. Obrecht. VIII, 299 pages. 1986.

Vol. 1224: Nonlinear Diffusion Problems, Montecatini Terme 1985. Seminar. Edited by A. Fasano and M. Primicerio. VIII, 188 pages. 1986.

Vol. 1225: Inverse Problems, Montecatini Terme 1986. Seminar. Edited by G. Talenti. VIII, 204 pages. 1986.

Vol. 1226: A. Buium, Differential Function Fields and Moduli of Algebraic Varieties. IX, 146 pages. 1986.

Vol. 1227: H. Helson, The Spectral Theorem. VI, 104 pages. 1986.

Vol. 1228: Multigrid Methods II. Proceedings, 1985. Edited by W. Hackbusch and U. Trottenberg. VI, 336 pages. 1986.

Vol. 1229: O. Bratteli, Derivations, Dissipations and Group Actions on C*-algebras. IV, 277 pages. 1986.

Vol. 1230: Numerical Analysis. Proceedings, 1984. Edited by J.-P. Hennart. X, 234 pages. 1986.

Vol. 1231: E.-U. Gekeler, Drinfeld Modular Curves. XIV, 107 pages. 1986.

Vol. 1232: P.C. Schuur, Asymptotic Analysis of Soliton Problems. VIII, 180 pages. 1986.

Vol. 1233: Stability Problems for Stochastic Models. Proceedings, 1985. Edited by V.V. Kalashnikov, B. Penkov and V.M. Zolotarev. VI, 223 pages. 1986.

Vol. 1234: Combinatoire énumérative. Proceedings, 1985. Edité par G. Labelle et P. Leroux. XIV, 387 pages. 1986.

Vol. 1235: Séminaire de Théorie du Potentiel, Paris, No. 8. Directeurs: M. Brelot, G. Choquet et J. Deny. Rédacteurs: F. Hirsch et G. Mokobodzki. III, 209 pages. 1987.

Vol. 1236: Stochastic Partial Differential Equations and Applications. Proceedings, 1985. Edited by G. Da Prato and L. Tubaro. V, 257 pages. 1987.

Vol. 1237: Rational Approximation and its Applications in Mathematics and Physics. Proceedings, 1985. Edited by J. Gilewicz, M. Pindor and W. Siemaszko. XII, 350 pages. 1987.

Vol. 1238: M. Holz, K.-P. Podewski and K. Steffens, Injective Choice Functions. VI, 183 pages. 1987.

Vol. 1239: P. Vojta, Diophantine Approximations and Value Distribution Theory. X, 132 pages. 1987.

Vol. 1240: Number Theory, New York 1984–85. Seminar. Edited by D.V. Chudnovsky, G.V. Chudnovsky, H. Cohn and M.B. Nathanson. V, 324 pages. 1987.

Vol. 1241: L. Gårding, Singularities in Linear Wave Propagation. III, 125 pages. 1987.

Vol. 1242: Functional Analysis II, with Contributions by J. Hoffmann-Jørgensen et al. Edited by S. Kurepa, H. Kraljević and D. Butković. VII, 432 pages. 1987.

Vol. 1243: Non Commutative Harmonic Analysis and Lie Groups. Proceedings, 1985. Edited by J. Carmona, P. Delorme and M. Vergne. V, 309 pages. 1987.

Vol. 1244: W. Müller, Manifolds with Cusps of Rank One. XI, 158 pages. 1987.

Vol. 1245: S. Rallis, L-Functions and the Oscillator Representation. XVI, 239 pages. 1987.

Vol. 1246: Hodge Theory. Proceedings, 1985. Edited by E. Cattani, F. Guillén, A. Kaplan and F. Puerta. VII, 175 pages. 1987.

Vol. 1247: Séminaire de Probabilités XXI. Proceedings. Edité par J. Azéma, P.A. Meyer et M. Yor. IV, 579 pages. 1987.

Vol. 1248: Nonlinear Semigroups, Partial Differential Equations and Attractors. Proceedings, 1985. Edited by T.L. Gill and W.W. Zachary. IX, 185 pages. 1987.

Vol. 1249: I. van den Berg, Nonstandard Asymptotic Analysis. IX, 187 pages. 1987.

Vol. 1250: Stochastic Processes – Mathematics and Physics II. Proceedings 1985. Edited by S. Albeverio, Ph. Blanchard and L. Streit. VI, 359 pages. 1987.

Vol. 1251: Differential Geometric Methods in Mathematical Physics. Proceedings, 1985. Edited by P.L. García and A. Pérez-Rendón. VII, 300 pages. 1987.

Vol. 1252: T. Kaise, Représentations de Weil et GL_2 Algèbres de division et GL_n. VII, 203 pages. 1987.

Vol. 1253: J. Fischer, An Approach to the Selberg Trace Formula via the Selberg Zeta-Function. III, 184 pages. 1987.

Vol. 1254: S. Gelbart, I. Piatetski-Shapiro, S. Rallis. Explicit Constructions of Automorphic L-Functions. VI, 152 pages. 1987.

Vol. 1255: Differential Geometry and Differential Equations. Proceedings, 1985. Edited by C. Gu, M. Berger and R.L. Bryant. XII, 243 pages. 1987.

Vol. 1256: Pseudo-Differential Operators. Proceedings, 1986. Edited by H.O. Cordes, B. Gramsch and H. Widom. X, 479 pages. 1987.

Vol. 1257: X. Wang, On the C*-Algebras of Foliations in the Plane. V, 165 pages. 1987.

Vol. 1258: J. Weidmann, Spectral Theory of Ordinary Differential Operators. VI, 303 pages. 1987.